云计算技术与应用专业系列教材

# Java Web
# 云应用开发

郑子伟 屈毅 主编
何福南 谢涛 陈张荣 副主编
李梅芳 金忠伟 主审



人 民 邮 电 出 版 社
北 京

**图书在版编目（CIP）数据**

Java Web云应用开发 / 郑子伟，屈毅主编. -- 北京：人民邮电出版社，2017.12
云计算技术与应用专业系列教材
ISBN 978-7-115-46234-3

Ⅰ. ①J… Ⅱ. ①郑… ②屈… Ⅲ. ①JAVA语言－程序设计－教材 Ⅳ. ①TP312.8

中国版本图书馆CIP数据核字(2017)第236127号

## 内 容 提 要

本书以开源的 OpenStack 技术为基础，按照“任务驱动、能力递进”的思路，以“Web 云网盘项目”的开发为主线，全面介绍了使用 Java Web 技术开发 Web 云网盘系统的方法和主要步骤。本书将 Web 云网盘的项目分解为 Web 云网盘概要设计、开发环境搭建、JavaEE 基础知识、云存储 OpenStack Swift 服务构建、开发登录注册模块、开发文件列表模块、开发文件操作模块、开发功能扩展模块、部署发布 9 个子项目，在每一个子项目中，首先提出任务目标，然后介绍任务的实现步骤，并对项目涉及的技术背景进行详细说明，最后给出每一个项目的实现效果。本书还提供了项目相关的源码和课件，以方便读者自学。学习本课程的读者需要初步了解 OpenStack 云存储 Swift，并具有 Java 语言基础。

本书可以作为高职高专云计算技术与应用专业的基础核心课程，以及计算机相关专业云计算选修课程的教材，也可以作为云计算基础入门的培训教材，并适合从事云计算开发、云计算运维、云计算销售技术支持的专业人员和广大计算机爱好者自学使用。

◆ 主　　编　郑子伟　屈　毅
副 主 编　何福南　谢　涛　陈张荣
主　　审　李梅芳　金忠伟
责任编辑　左仲海
责任印制　马振武

◆ 人民邮电出版社出版发行　　北京市丰台区成寿寺路 11 号
邮编　100164　　电子邮件　315@ptpress.com.cn
网址　http://www.ptpress.com.cn
大厂聚鑫印刷有限责任公司印刷

◆ 开本：787×1092　1/16
印张：14.25　　　　2017 年 12 月第 1 版
字数：334 千字　　　2017 年 12 月河北第 1 次印刷

定价：39.80 元

读者服务热线：(010)81055256　印装质量热线：(010)81055316
反盗版热线：(010)81055315
广告经营许可证：京东工商广登字 20170147 号

# 前 言 FOREWORD

计算机技术经历了从大型主机、个人计算机、客户/服务器计算模式到今天的互联网计算模式的演变，尤其是互联网 Web 2.0 技术的应用，使得对计算能力的需求更多地依赖于通过互联网连接的远程服务器资源。作为资源的提供者，需要具备超高的计算性能、海量的数据存储、网络通信能力和随时的扩展能力。在多种应用需求的推动下催生了虚拟化技术和云计算技术。当今，云计算技术已经成为信息技术应用服务平台、云存储技术、大数据分析、互联网+等技术的基础支撑，在信息技术发展过程中起着平台支撑的作用。

云计算是推动信息技术实现按需供给、促进信息技术和数据资源充分利用的全新业态，是信息化发展的重大变革和必然趋势。推动和发展云计算技术，有利于分享信息知识和创新资源，降低全社会创业成本，培育形成新产业和新消费热点，对稳增长、调结构、惠民生和建设创新型国家具有重要意义。

为适应高职院校对云计算技术与应用专业教学的需求，在“云计算技术与应用专业教材编审委员会”的组织和指导下，将陆续推出系列专业教材。本书就是在此背景下，由南京第五十五所技术开发有限公司、江苏一道云科技发展有限公司共同编写，是校企产教融合后的实践产物。本书是基于开源的 OpenStack 云计算架构，针对高职高专云计算专业或相关专业的云计算应用与开发课程，以 Web 云网盘项目的开发为主线，按照“任务驱动、能力递进”的思路编写而成的。本书将 Web 云网盘项目的开发分解为若干子项目，对每一个子项目，首先提出任务目标，然后详细描述任务的实现步骤，并对项目涉及的技术原理进行详细介绍，最后给出每一个项目的实现效果。全书力争做到基础知识介绍有针对性，任务目标操作具体化。Web 云网盘项目以云存储 OpenStack Swift 服务为基础，使用业内主流的 Java Web 技术开发，项目功能全面、系统扩展性强，具有良好的实用性。

本书的参考学时为 54 ~ 60 学时，建议采用理论实践一体化教学模式，各项目的参考学时见学时分配表。

学时分配表

| 项　　目 | 课 程 内 容 | 学　　时 |
| --- | --- | --- |
| 项目 1 | Web 云网盘软件概要设计 | 4 |
| 项目 2 | 开发环境搭建 | 4 |
| 项目 3 | JavaEE 基础知识 | 4 |
| 项目 4 | 云存储 OpenStack Swift 服务构建 | 6 |
| 项目 5 | 开发登录注册模块 | 6 |
| 项目 6 | 开发文件列表模块 | 8 |
| 项目 7 | 开发文件操作模块 | 8 |
| 项目 8 | 开发功能扩展模块 | 10 |
| 项目 9 | 部署发布 | 2 |
|  | 课程考评 | 2 |
| 课时总计 |  | 54 |

本教材由郑子伟、屈毅主编，何福南、谢涛、陈张荣副主编，李梅芳、金忠伟主审。本书配套的资源包、运行脚本、电子教案等，可登录 http://www.1daoyun.com 下载。

对于本书，虽然编者已尽可能做到更好，但由于搭建环境的复杂性，书中疏漏和不足之处在所难免，殷切希望广大读者批评指正。同时，恳请读者一旦发现错误，于百忙之中及时与编者联系，以便尽快更正，编者将不胜感激，E-mail：books@cetc55.net。

编　者

2017 年 8 月

目 录

CONTENTS

# 项目 1 Web 云网盘软件概要设计

## 单元介绍

本单元主要介绍 Web 云网盘软件的主要功能和技术架构。

## 学习任务

本单元主要完成以下学习目标：

- 了解云存储的基础知识；
- 了解 OpenStack 的基础知识；
- 了解项目的界面原型图设计；
- 了解项目的技术框架选型。

## 任务 1.1 了解云存储的基本知识

### 1.1.1 相关知识

云存储是一种新兴的网络存储技术，是指通过集群应用、网络技术或分布式文件系统等功能，将网络中大量各种不同类型的存储设备通过应用软件集合起来协同工作，对外提供数据存储和业务访问的系统。使用者可以在任何时间、任何地方，通过任何可联网的装置连接到云上，从而方便地存取数据。

对象存储是实现云存储的一种主要方式。对象存储是网络提供的基于互联网的简单对象存储服务，提供简单易用的 RESTful API 接口，使客户在任何时间、任何地点都能通过互联网访问对象存储的数据。通过对象存储服务，客户不用担心数据丢失、数据安全、数据存储空间等问题，把更多精力聚焦在如何利用数据创新等业务上。

Swift 最初是由 Rackspace 公司开发的高可用分布式对象存储服务，并于 2010 年贡献给 OpenStack 开源社区作为其最初的核心子项目之一，为其 Nova 子项目提供虚机镜像存储服务。Swift 构筑在廉价的标准硬件存储基础设施之上，无需采用 RAID（磁盘冗余阵列），通过在软件层面引入一致性散列技术和数据冗余性，牺牲一定程度的数据一致性来达到高可用性和可伸缩性，支持多租户模式、容器和对象读写操作，适合解决互联网的应用场景下非结构化数据存储问题。

在 OpenStack 中，Swift 主要用于存储虚拟机镜像和 Glance 的后端存储。在实际运用中，

Swift 的典型运用是网盘系统，存储的数据类型大多为图片、邮件、视频、存储备份等静态资源。

Swift 不能像传统文件系统那样进行挂载和访问，只能通过 RESTful API 接口访问数据。Swift 不同于传统文件系统和实时数据存储系统，它适用于存储和获取一些静态的永久性数据，并在需要的时候进行更新。

OpenStack Swift 作为稳定和高可用的开源对象存储被很多企业作为商业化部署，如 Rackspace 公司通过结合 Swift 和 Nova 提供 IaaS 云服务，微软的 SharePoint、新浪的 App Engine、韩国电信的 Ucloud Storage 服务均使用了 OpenStack Swift 技术。有理由相信，因为其完全的开放性、广泛的用户群和社区贡献者，Swift 可能会成为云存储的开放标准。

### 1.1.2 实现步骤

#### 1. 主要功能设计

Web 云网盘软件（以下简称网盘）是一个云存储系统。该软件可以为每个用户开辟独立的存储空间，使用户随时随地访问和管理文档资产。该软件使用 OpenStack Swift 服务实现，实现思路如下。

（1）在 OpenStack 创建一个租户（Swift），租户下包括多个用户。租户创建云存储空间 100GB。

（2）每个用户支持注册和登录功能。如果是新用户，进行注册，同时注册一个租户，以及租户下的用户。如果是已有的用户，可以进行登录及管理登录。

（3）每个用户登录时，如果没有容器，默认创建两个容器，包括 Container 目录和 Recycle 回收站。其中 Container 目录用于存放常规数据，Recycle 回收站用于存放用户删除的数据。

（4）为每个用户分配单独的存储空间，共享存储空间为所有用户共有。

（5）网盘提供常见的文件操作功能。

网盘的主要功能见表 1-1。

表 1-1 网盘的主要功能

| 编号 | 分类 | 操作 | 说　明 | 分类 |
|---|---|---|---|---|
| 1 | 查看 | 搜索 | 搜索文件或目录，一般支持通配符 | 基本功能 |
| 2 | | 下载 | 下载一个文件到本地 | 基本功能 |
| 3 | | 打开文件 | 打开文件，在线预览 | 基本功能 |
| 4 | | 进入目录 | 进入子目录 | 基本功能 |
| 5 | | 详情 | 查看云存储空间的详细信息，包括文件数量、文件分类数量、存储空间总容量等数据 | 扩展功能 |
| 6 | | 全选 | 能够对列表中的全部文件进行一键全选操作 | 综合实战 |
| 7 | | 缩略图展示 | 文件以网格缩略图的形式展示 | 综合实战 |
| 8 | | 排序 | 按照时间、大小、名称进行排序 | 综合实战 |

续表

| 编号 | 分类 | 操作 | 说 明 | 分类 |
|---|---|---|---|---|
| 9 | 增加 | 上传文件 | 上传文件与某文件重名将覆盖原文件，一旦覆盖不能还原 | 基本功能 |
| 10 | 删除 | 删除文件 | 删除一个文件或多个文件到回收站 | 基本功能 |
| 11 | | 删除文件夹 | 删除一个或多个文件夹到回收站 | 扩展功能 |
| 12 | | 清空回收站 | 把回收站清空 | 基本功能 |
| 13 | | 还原文件 | 将回收站中的文件或文件夹还原到网盘中 | 综合实战 |
| 14 | 修改 | 修改文件名称 | 对指定文件进行改名 | 基本功能 |
| 15 | | 修改目录名称 | 对指定目录进行改名 | 基本功能 |
| 16 | 分享 | 分享 | 选择一个文件，分享整个数据给外部使用 | 扩展功能 |
| 17 | 群分享 | 群分享 | 建立一个群，群里的用户可以分享文件 | 综合实战 |

### 2. 原型界面设计

分析了用户的功能操作流程后，就可以对项目进行界面设计了。下面介绍本项目的主要操作界面。

（1）注册登录界面

登录云网盘界面如图 1-1 所示。

图 1-1 登录界面

（2）文件管理界面

文件管理和操作界面如图 1-2 所示。

图 1-2　文件管理和操作界面

（3）文件分享界面

文件分享界面如图 1-3 所示。

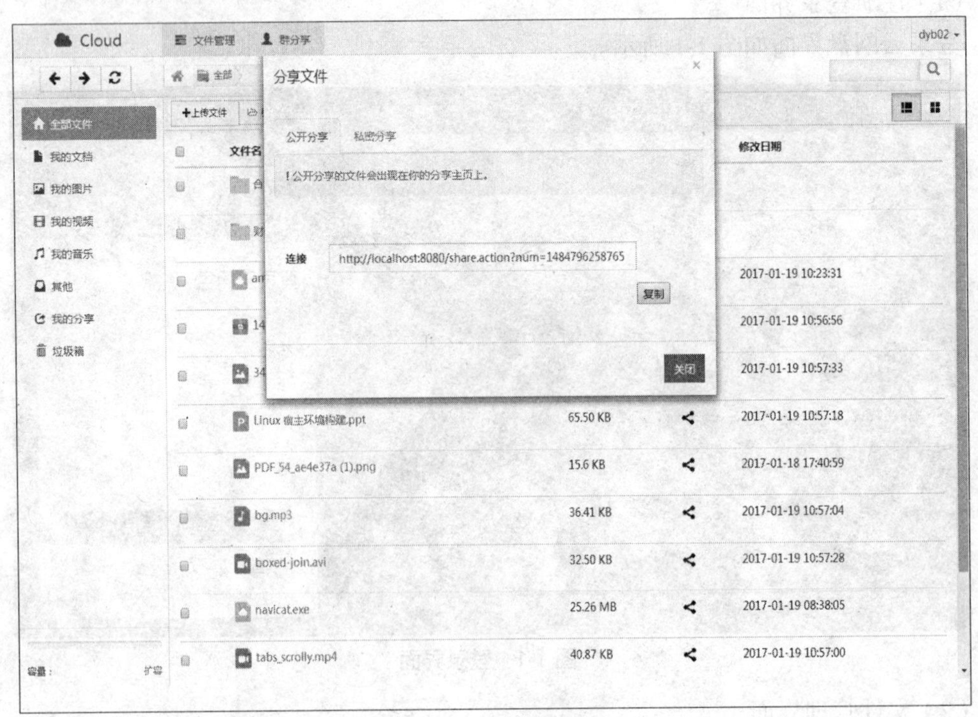


图 1-3　文件分享界面

群文件分享界面如图 1-4 所示。

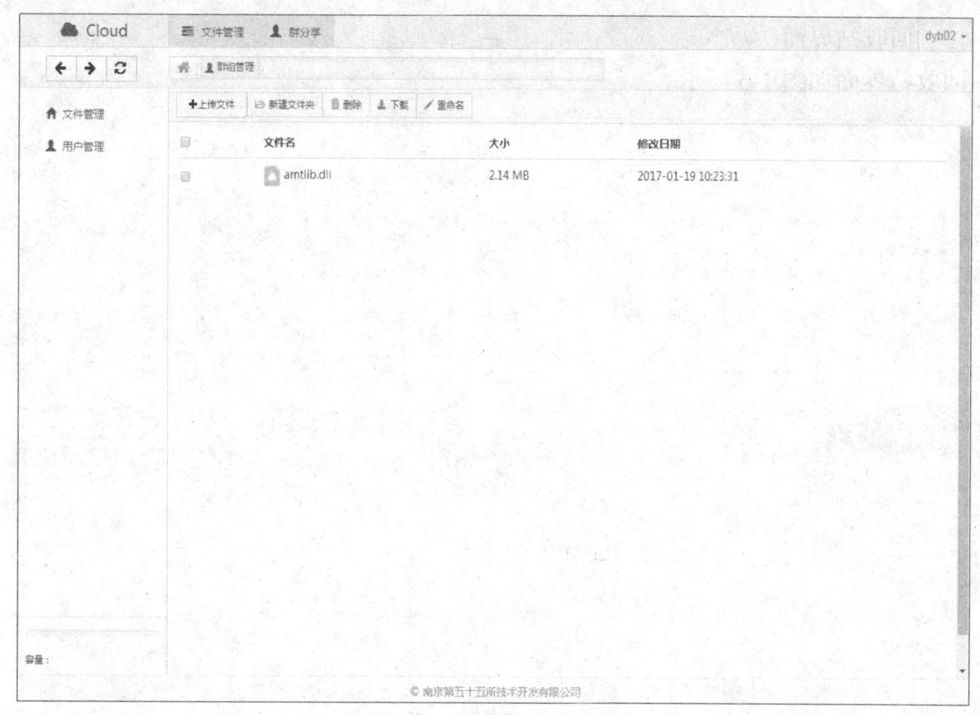


图 1-4　群文件分享界面

（4）文件预览界面

文件预览界面如图 1-5 所示。

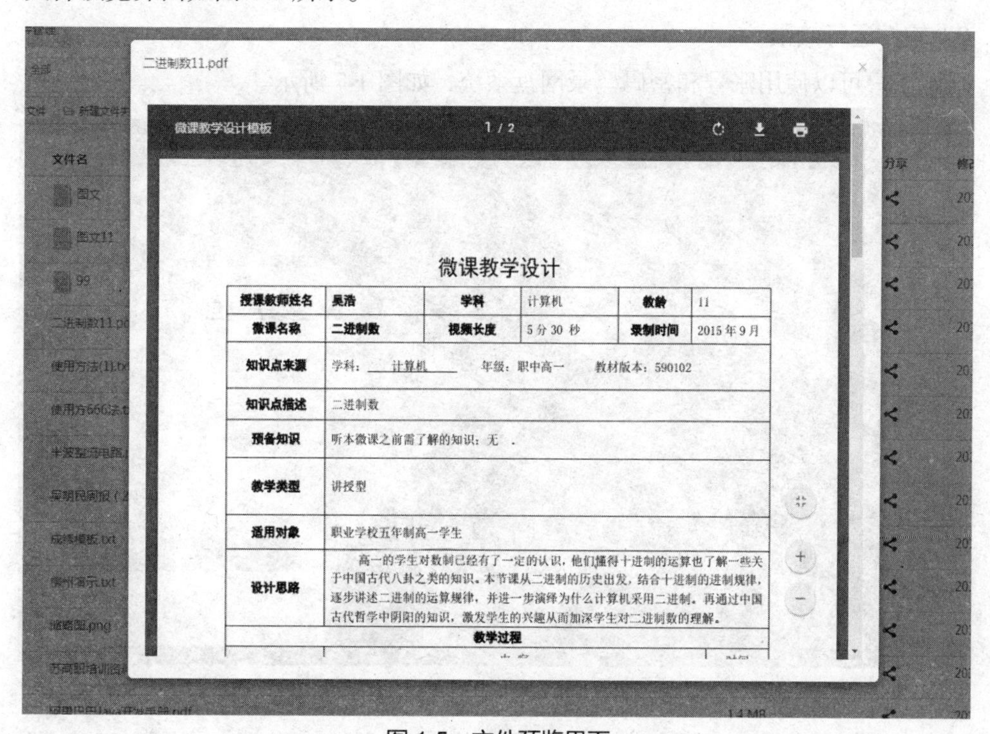


图 1-5　文件预览界面

（5）回收站界面

回收站界面如图 1-6 所示。

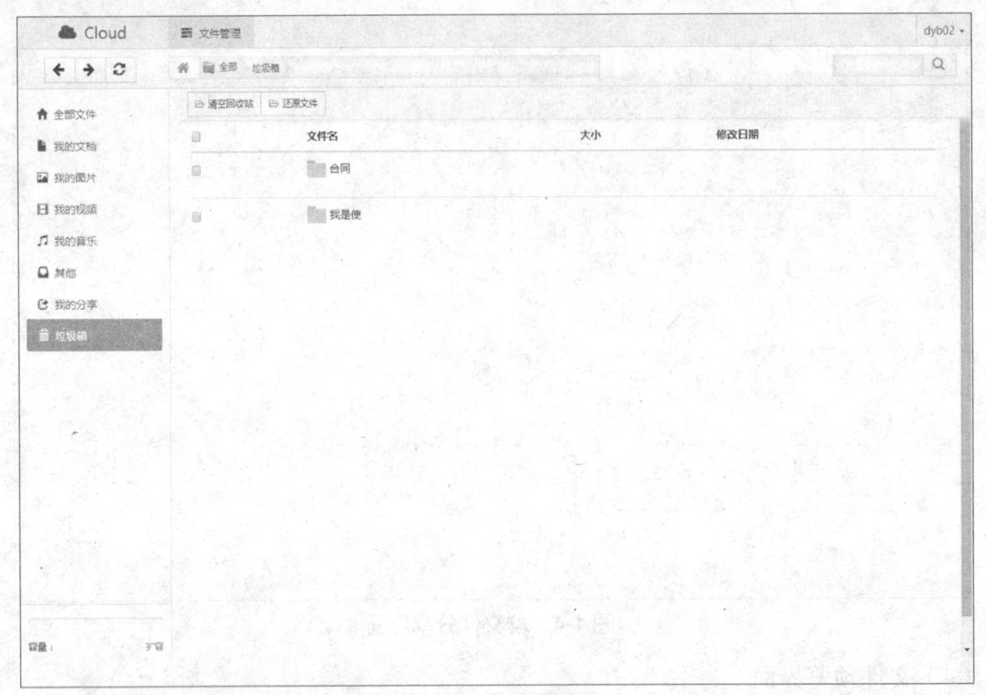


图 1-6　回收站界面

3．软件运行效果

（1）用户可以使用账号和密码登录网盘系统，如图 1-7 所示。

图 1-7　登录界面

登录成功后，系统显示该用户的所有文件列表，如图 1-8 所示。

图 1-8　所有文件列表

（2）通过左侧的导航菜单可以选择显示某类型的文件列表，例如，可以选择“我的图片”选项，网盘系统则显示该用户的所有图片文件。图片列表有两种显示方式：列表视图和网格视图。其中列表视图为默认显示视图，显示效果如图 1-9 所示。

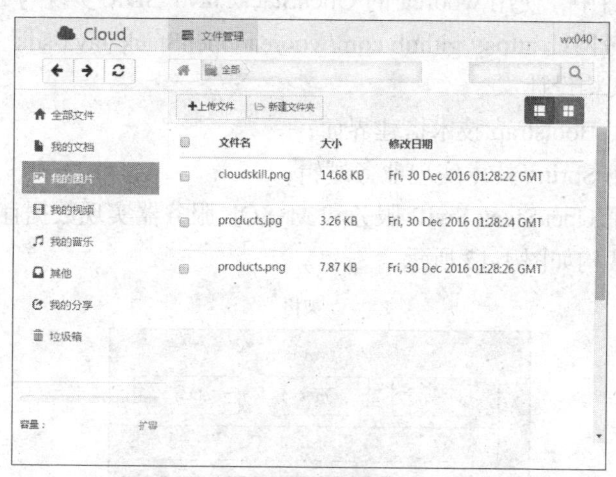


图 1-9　列表视图

切换到网格视图，显示缩略图效果，如图 1-10 所示。

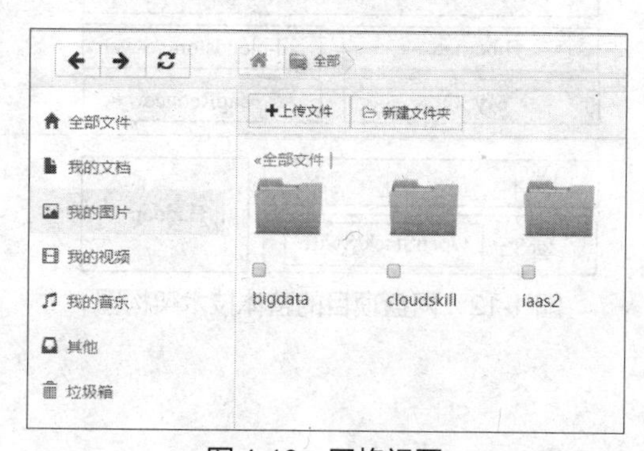


图 1-10　网格视图

（3）在列表视图中，中间的一排按钮是文件操作菜单，也是系统的主要功能入口，如图 1-11 所示。

图 1-11　文件操作功能菜单

4. 技术架构设计

OpenStack 经过多年的发展，参与这个开源云计算项目的公司和人员越来越多，OpenStack 社区也越来越活跃，对 OpenStack 的支持会越来越丰富。目前 OpenStack 社区为 OpenStack 提供了多种语言的 SDK，包括 Python、Java、PHP 等语言，后续还会有更多其他语言的 SDK 出现。

基于 OpenStack 的开发，可以使用 OpenStack 本身的 API 开发，也可以使用第三方的 SDK 开发。在本项目中，使用 woorea 的 OpenStack Java SDK 实现与 Swift 通信的功能，woorea 的使用可参考网址 https://github.com/woorea/OpenStack-Java-sdk。

本项目为三层架构设计：

（1）显示层使用 Bootstrap 技术搭建界面；

（2）业务层使用 Spring 技术实现业务逻辑；

（3）存储层使用 OpenStack Swift 服务和 MySQL 服务器实现数据存储和管理。

本项目的整体架构如图 1-12 所示。

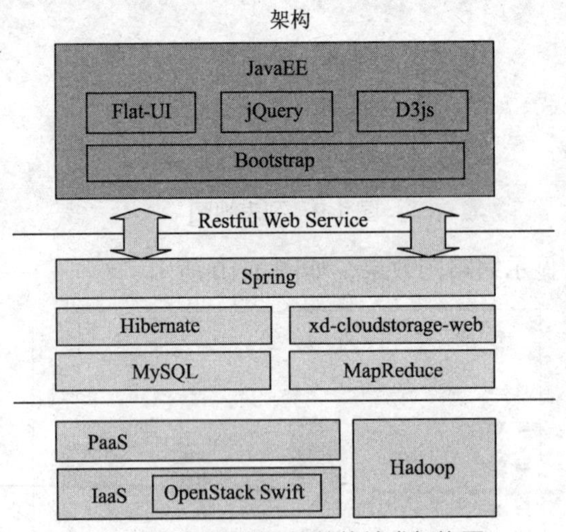


图 1-12　网盘项目的整体技术架构图

# 项目 2 开发环境搭建

## 单元介绍

本单元读者需要学习如何安装部署 Web 项目的运行环境，包括安装 JDK、安装开发工具 Eclipse、安装运行 Web 服务器 Tomcat，以及安装 MySQL 数据库。

## 学习任务

本单元主要完成以下学习目标：

- 掌握 JDK 的安装与配置；
- 掌握 Tomcat 的安装与配置；
- 掌握 Eclipse 的安装与配置；
- 掌握 MySQL 的安装与配置；
- 掌握 Eclipse 创建 Web 工程的基本步骤。

根据以上学习目标，本项目将分解为以下 5 个任务，具体见表 2-1。

表 2-1 任务分解表

| 任务序号 | 任务内容 | 验证方式 |
| --- | --- | --- |
| 任务 2.1 | 安装配置 JDK | 执行获取 Java 版本命令 |
| 任务 2.2 | 安装配置 Tomcat | 正常执行终端模拟器 |
| 任务 2.3 | 安装配置 Eclipse | 新建 Web 工程 |
| 任务 2.4 | 安装配置 MySQL | 执行获取 MySQL 版本命令 |
| 任务 2.5 | 新建 HelloWorld 程序 | Hello World 程序可以正常运行 |

## 任务 2.1 安装配置 JDK

### 2.1.1 相关知识

JDK（Java Development Kit）称为 Java 开发包或 Java 开发工具，是一个编写 Java 的 Applet 小程序和应用程序的程序开发环境。JDK 是整个 Java 的核心，包括了 Java 运行环境（Java Runtime Environment）、Java 工具和 Java 的核心类库（Java API）。主流的 JDK 是

Sun 公司发布的 JDK，除了 Sun 之外，还有很多公司和组织都开发了自己的 JDK。例如，IBM 公司开发的 JDK，BEA 公司的 Jrocket，还有 GNU 组织开发的 JDK。

另外，可以把 Java API 类库中的 Java SE API 子集和 Java 虚拟机这两部分统称为 JRE（Java Runtime Environment），JRE 是支持 Java 程序运行的标准环境。

需要说明的是，JRE 是运行环境，JDK 是开发环境，因此编写 Java 程序时需要 JDK，而运行 Java 程序的时候就需要 JRE。另外，JDK 里面已经包含了 JRE，因此只要安装了 JDK，就可以编辑 Java 程序，也可以正常运行 Java 程序。但由于 JDK 包含了许多与运行无关的内容，占用的空间较大，因此运行普通的 Java 程序无需安装 JDK，而只需要安装 JRE 即可。

## 2.1.2 实现步骤

（1）单击 JDK 安装程序，安装向导界面，如图 2-1 所示。

（2）设置安装路径，如图 2-2 所示。单击“更改”按钮，可以修改安装路径。

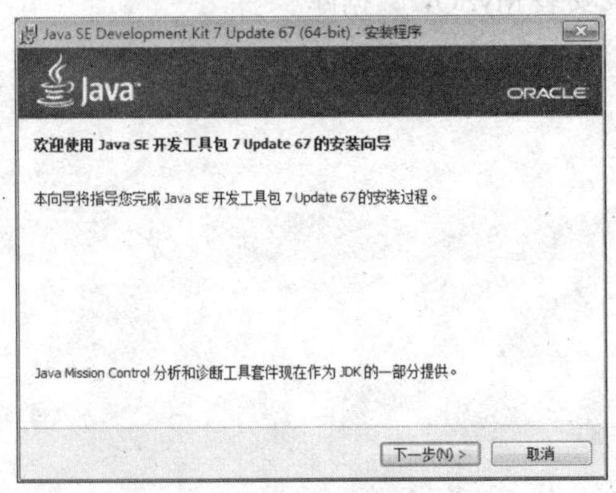


图 2-1 开始安装 JDK

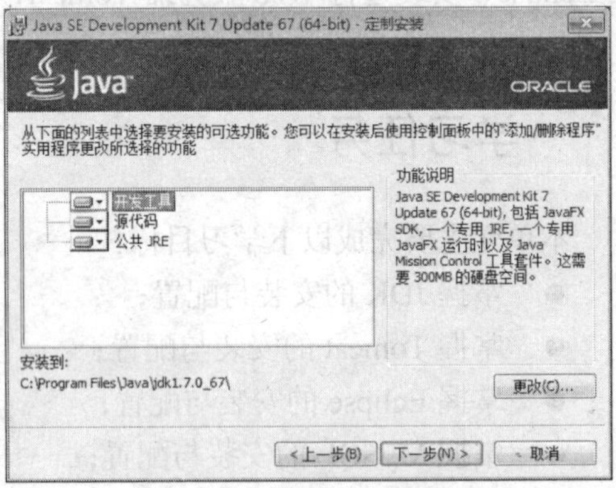

图 2-2 设置安装路径

（3）安装 JRE。在安装完 JDK 后，会弹出安装 JRE 窗口的界面（在这里要注意修改安装路径必须和 JDK 在同一目录下，而不是安装在 JDK 目录下），如图 2-3 所示。

（4）验证安装是否成功。安装完成后需要验证 Java 环境是否安装成功。验证方式有多种，可以编写一个最简单的 Java 程序文件后编译执行它，也可以通过显示 Java 版本的命令方式进行验证。本书采用通过输入 java -version 验证安装是否成功。读者也可以通过这个命令检查本机所安装的 Java 环境的版本。具体命令如图 2-4 所示。

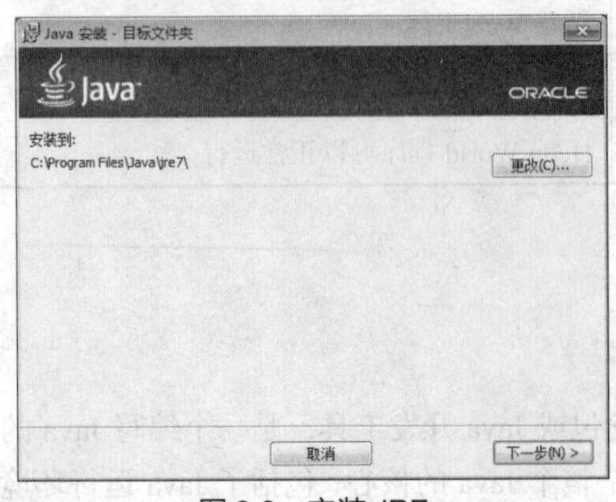


图 2-3 安装 JRE

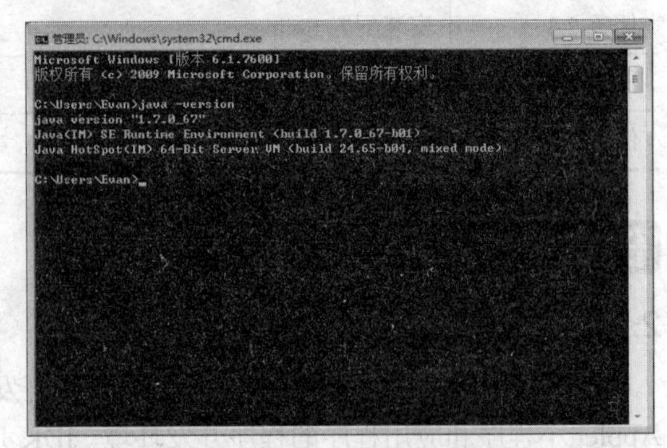


图 2-4 验证 Java 环境安装情况

## 任务 2.2 安装配置 Tomcat

### 2.2.1 相关知识

Tomcat 最初由 Sun 的软件构架师詹姆斯·邓肯·戴维森开发，后由 Sun 贡献给 Apache 软件基金会，目前是 Apache 软件基金会的顶级项目。Tomcat 是一个免费并且开源的 Web 服务器，支持 Servlet 和 JSP 规范。Tomcat 免费开源、技术先进、性能稳定，深受 Java 爱好者的喜爱并得到了部分软件开发商的认可，是目前比较流行的 Web 应用服务器。

### 2.2.2 实现步骤

#### 1. 下载安装 Tomcat

（1）从 Apache 网站（http://tomcat.apache.org/）下载 Tomcat，本书采用 Tomcat7.0 版本开发。

（2）将下载好的 Tomcat 压缩包解压至本地磁盘，如图 2-5 所示。

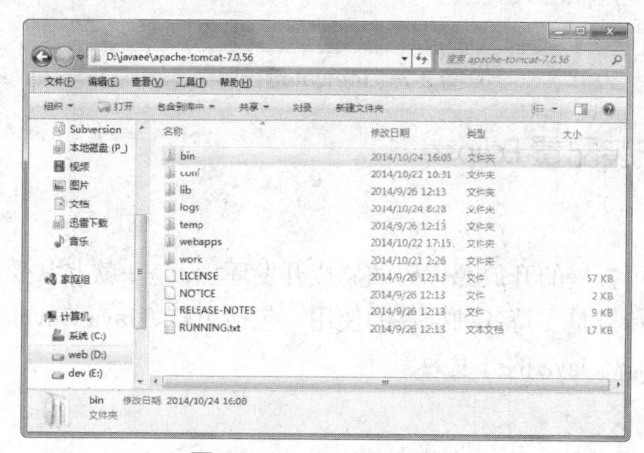


图 2-5 Tomcat 目录结构

#### 2. 运行测试 Tomcat

（1）在 Tomcat 文件夹下进入 bin 子文件夹，双击“startup.bat”文件，启动 Tomcat 服务器，如图 2-6 所示。

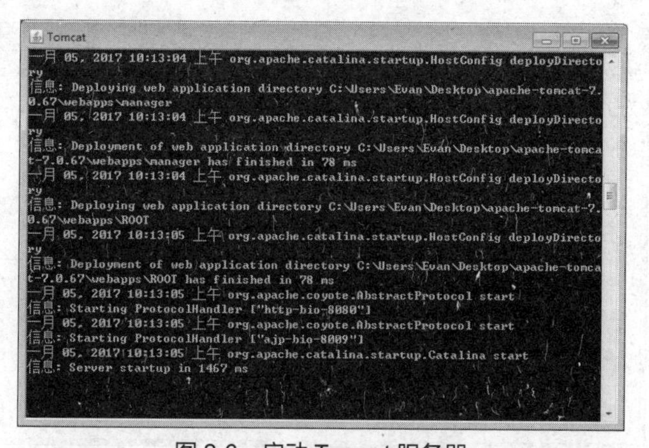


图 2-6 启动 Tomcat 服务器

（2）在浏览器的地址栏内输入 http://localhost:8080，如果 Tomcat 服务器部署成功，则显示如图 2-7 所示的界面。

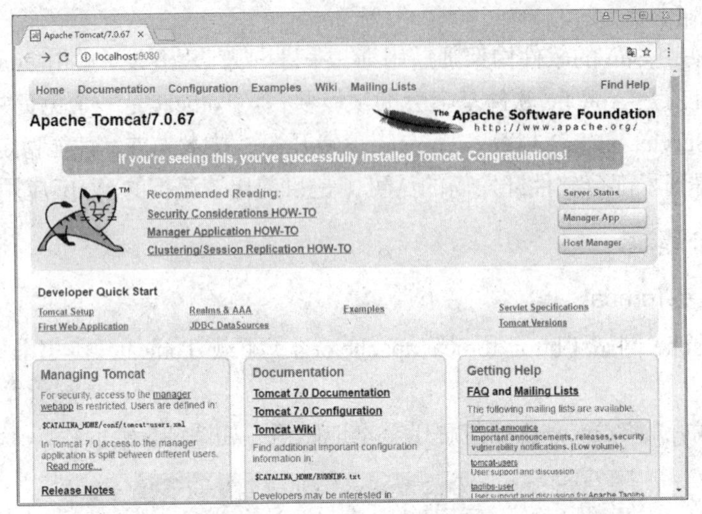

图 2-7　Tomcat 运行界面

## 任务 2.3　安装配置 Eclipse

### 2.3.1　相关知识

Eclipse 是一种可扩展的开放源代码的集成开发环境，是一款非常受欢迎的 Java 开发工具。Eclipse 的最大特点是支持各种插件的使用，包括 JDT（Java Development Tools），利用该插件可以大大提高 Java 的开发效率。

### 2.3.2　实现步骤

#### 1. 下载安装 Eclipse

读者可以从 Eclipse 的官网（http://www.eclipse.org）网站获取最新版本的 Eclipse，一般选择下载 Eclipse IDE for Java EE Developers 版本。

将下载好的 Eclipse 压缩包解压至电脑的本地磁盘，解压好的文件如图 2-8 所示。

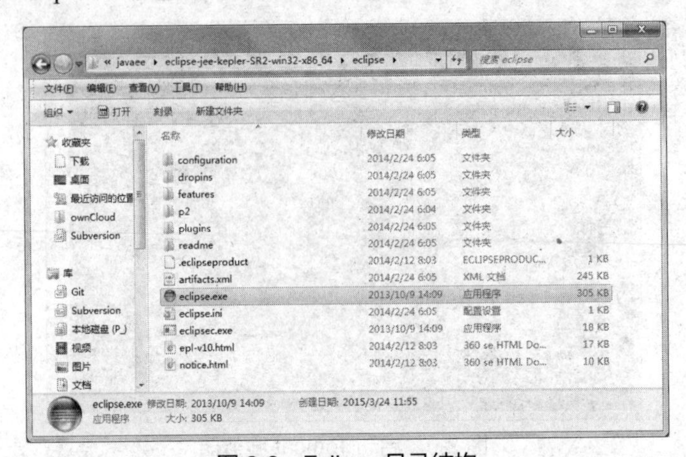

图 2-8　Eclipse 目录结构

2. 运行配置 Eclipse

（1）双击“eclipse.exe”文件启动 Eclipse，选择工作目录，注意该目录不要有中文信息，单击“OK”按钮，如图 2-9 所示。

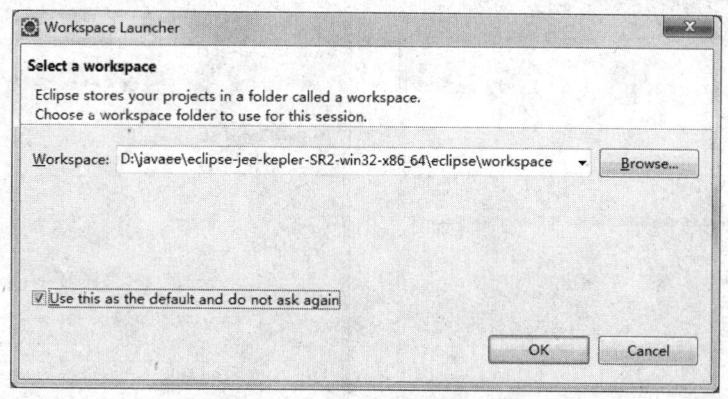


图 2-9　Eclipse 选择工作目录

（2）配置 Eclipse 的 Web 服务器。选择“Window”→“Preferences”→“Server”→“Runtime Environments”菜单命令，如图 2-10 所示。

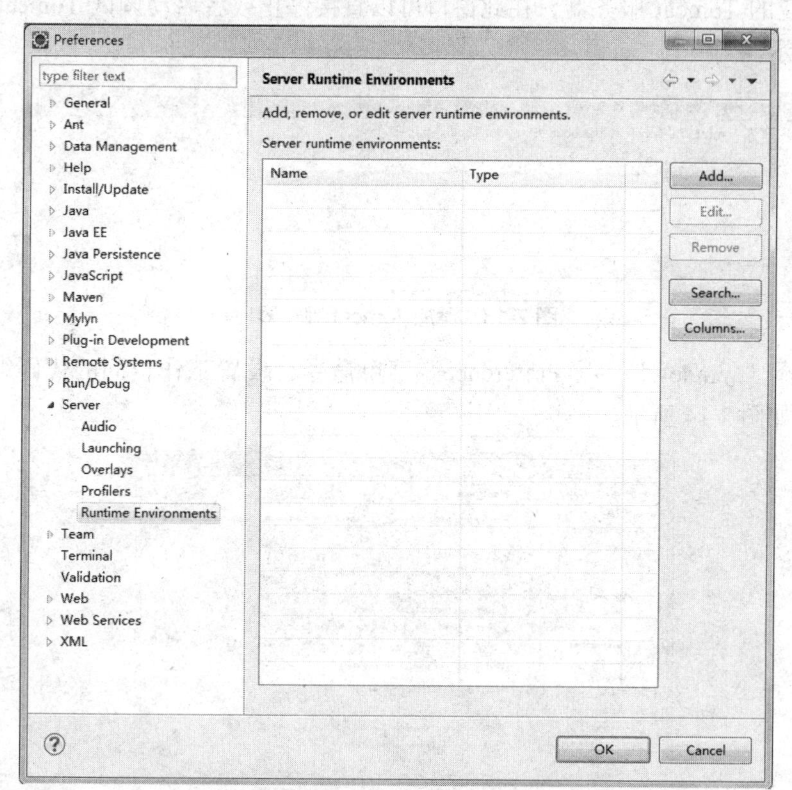


图 2-10　配置服务运行环境

（3）单击右侧“Add...”按钮，添加 Tomcat 服务器，如图 2-11 所示。

（4）选择 Apache Tomcat v7.0，单击“Next”按钮，设置 Tomcat 安装路径，单击“Finish”按钮完成配置，如图 2-12 所示。

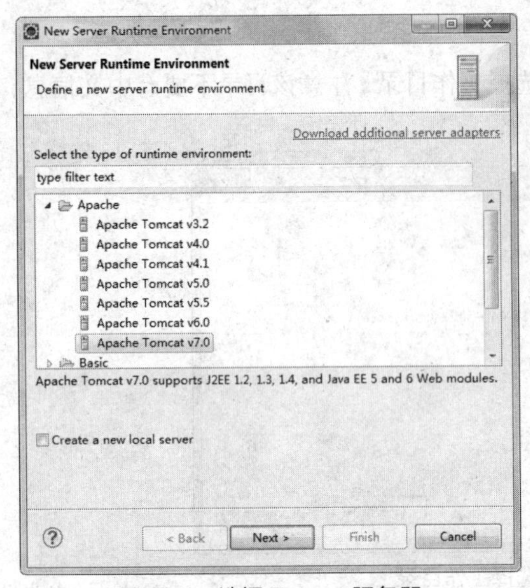


图 2-11　选择 Tomcat 服务器

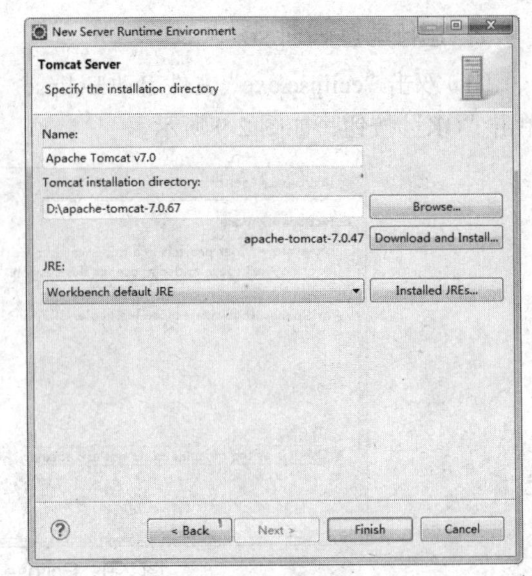


图 2-12　配置 Tomcat

（5）选择“Window”→“Show View”→“Servers”菜单命令，打开 Servers 窗口，添加刚才创建的 Tomcat 服务器，在此窗口可以直接停止或者启动调试 Tomcat 服务器，如图 2-13 所示。

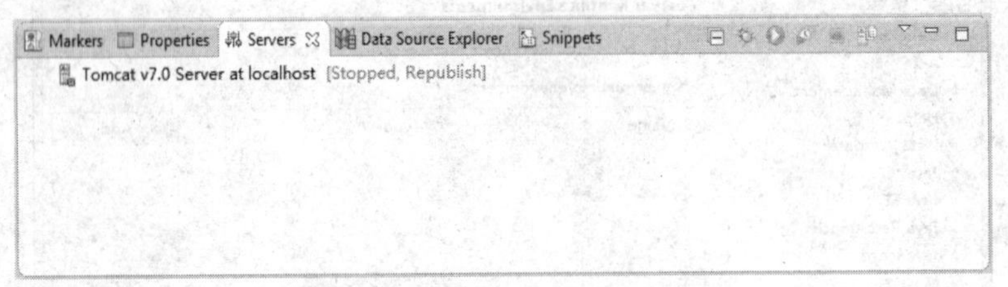


图 2-13　添加 Tomcat 服务器

（6）选择“Window”→“Preference”菜单命令，设置工作空间的代码编码方式为“UTF-8”，如图 2-14 所示。

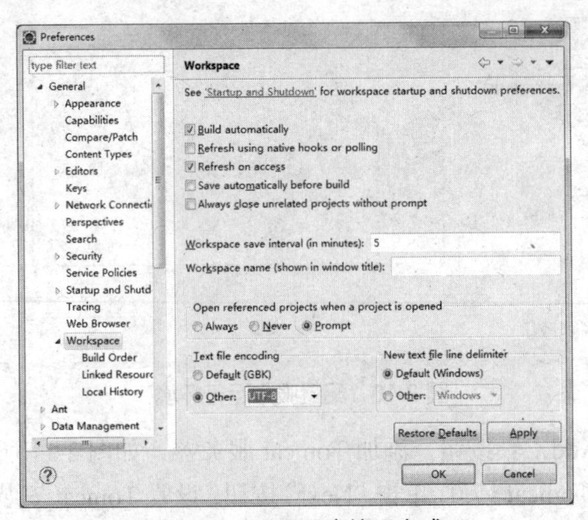


图 2-14　设置程序编码方式

## 任务 2.4　安装配置 MySQL

### 2.4.1　相关知识

MySQL 是一个关系型数据库管理系统，最早由瑞典 MySQL AB 公司开发，目前为 Oracle 公司的产品。MySQL 是业内流行的关系型数据库，MySQL 软件采用了双授权政策，它分为社区版和商业版，由于其体积小、速度快、开放源码等特点，一般中小型网站的开发都选择 MySQL 作为数据库。

### 2.4.2　实现步骤

1. 安装 MySQL 数据库

（1）双击打开 MySQL Server 5.0 安装程序“MySQL\Setup.exe”，单击“Next”按钮，如图 2-15 所示。

（2）在出现选择安装类型的窗口中，有“Typical（默认）”“Complete（完全）”“Custom（用户自定义）”三个选项，选择“Custom”选项，单击“Next”按钮，如图 2-16 所示。

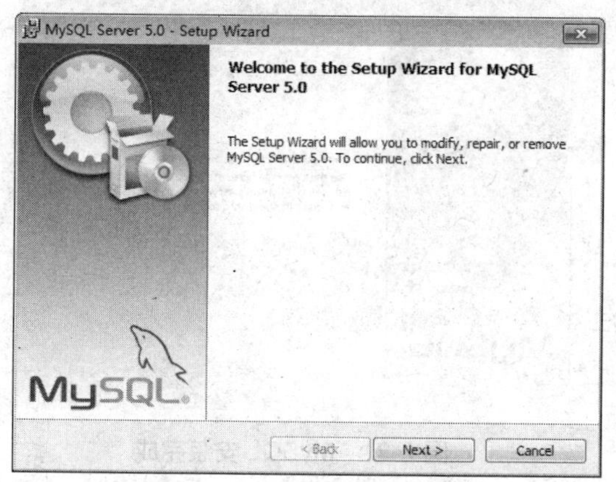


图 2-15　MySQL 安装向导

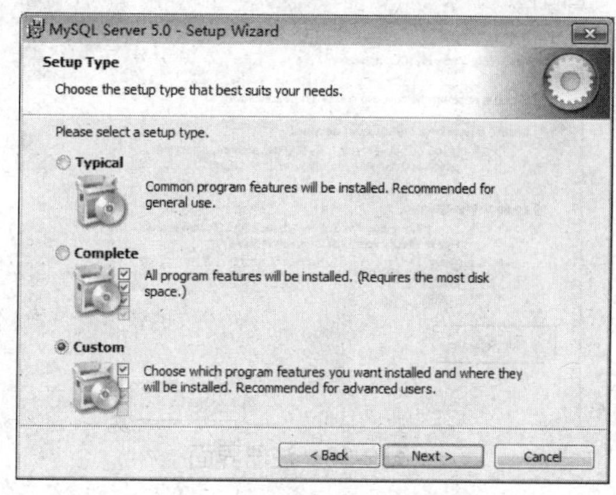


图 2-16　选择安装的类型

（3）在出现自定义安装界面中选择 MySQL 数据库的安装路径，本书设置的路径是“D:\Program File\MySQL”，在“Developer Components（开发者部分）”节点上单击，选择“This feature and all subfeatures will be installed on local hard drive.”，完成上述操作后单击“Next”按钮，如图 2-17 所示。

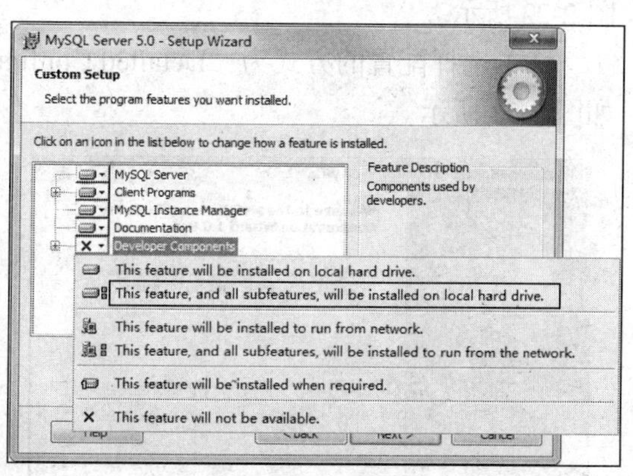


图 2-17　安装设置

（4）接下来进入准备安装的界面。首先确认刚才的设置是否正确，如果有误，单击“Back”按钮返回重新设置。如果之前的配置没有错误，单击“Install”按钮继续安装，如图 2-18 所示。

（5）单击“Install”按钮，出现正在安装的界面，如图 2-19 所示。

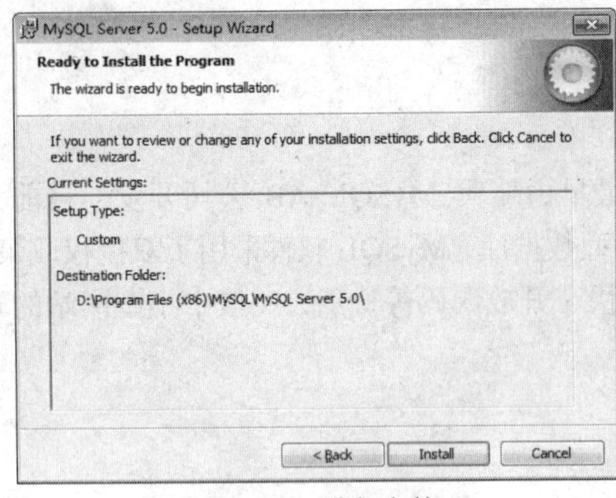


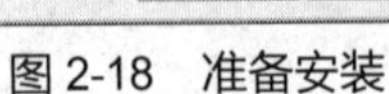


图 2-18　准备安装

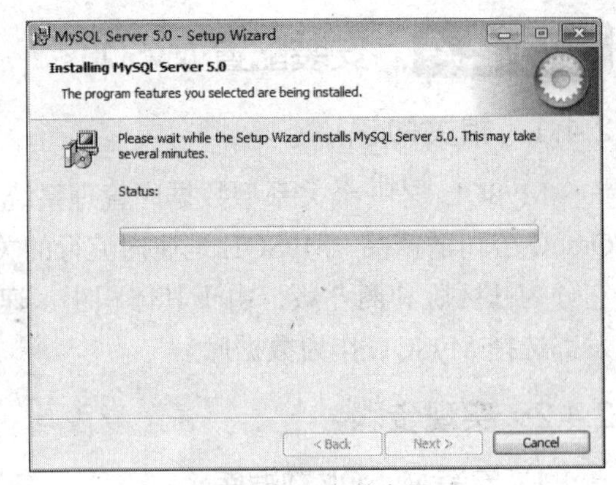


图 2-19　安装向导

（6）安装完成后出现注册界面，勾选“Skip Sign-Up”选项，单击“Next”按钮，如图 2-20 所示。

（7）单击“Finish”按钮，如图 2-21 所示。

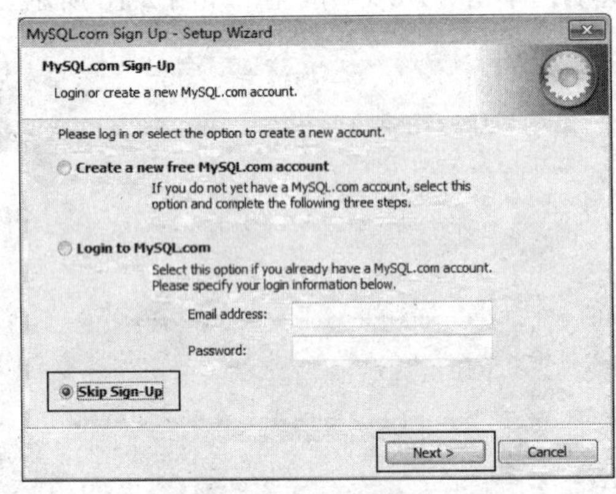


图 2-20　注册界面

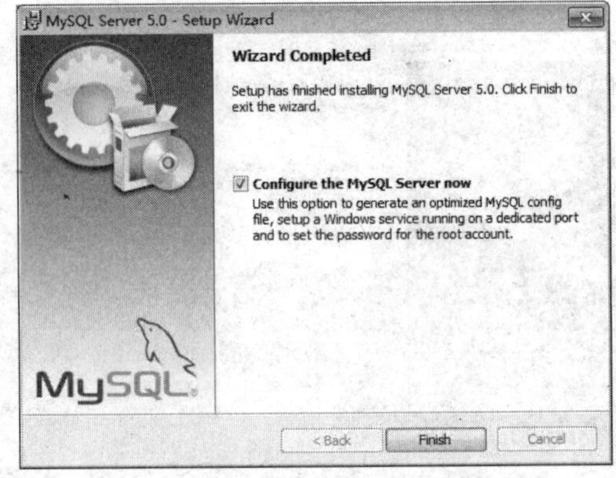


图 2-21　MySQL 安装完成

（8）MySQL 数据库安装完成之后，出现如下的配置界面向导，单击“Next”按钮，如图 2-22 所示。

（9）选择配置的方式为“Detailed Configuration（详细配置）”，单击“Next”按钮，如图 2-23 所示。

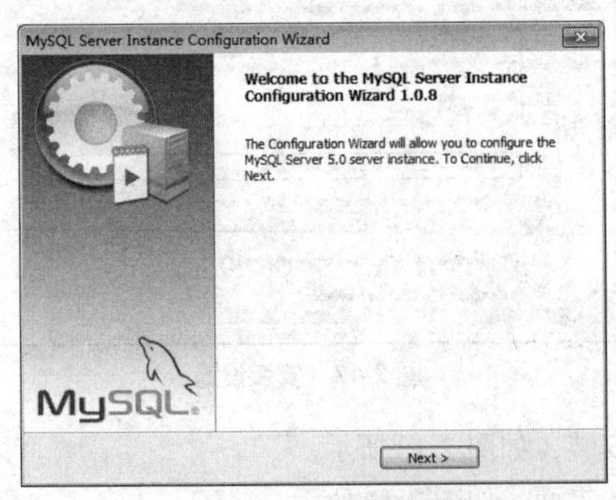


图 2-22　MySQL 配置参数向导

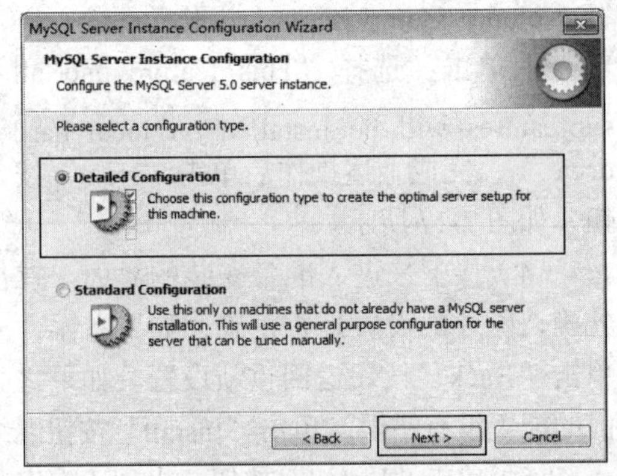


图 2-23　配置类型

（10）选择服务器的类型为“Developer Machine（开发测试类）”，单击“Next”按钮，如图 2-24 所示。

（11）设置 MySQL 数据库的用途，选择“Multifunctional Database（通用多功能型）”，单击“Next”按钮，如图 2-25 所示。

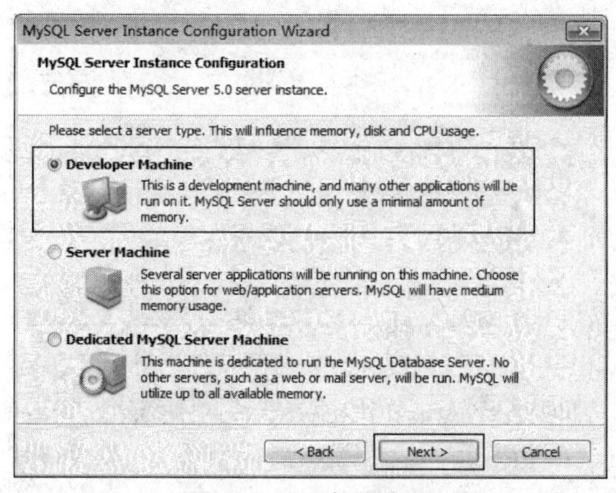


图 2-24　服务器类型

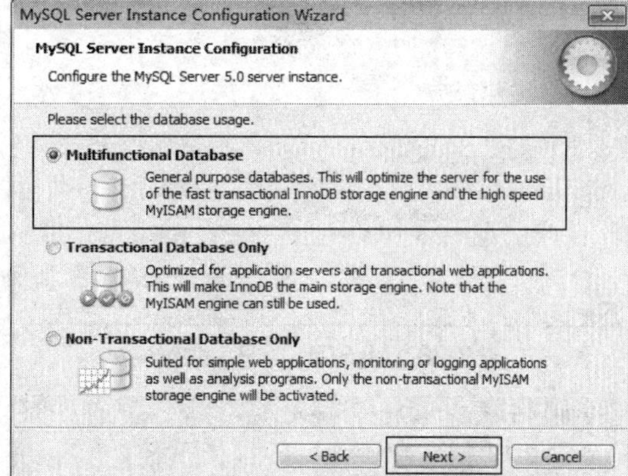


图 2-25　数据库用途

（12）对 InnoDB Tablespace 进行配置，选择 InnoDB 数据库文件存储位置，默认安装目录“Installation Path”，单击“Next”按钮，如图 2-26 所示。

（13）配置 MySQL 服务器的参数，选择“Manual Setting（手动设置）”，单击“Next”按钮，如图 2-27 所示。

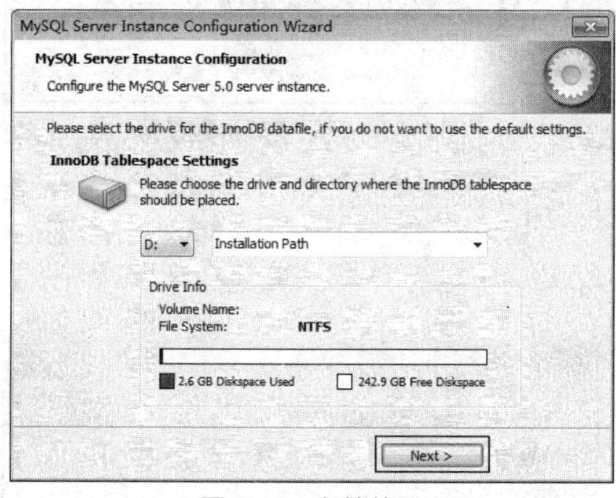


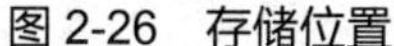

图 2-26　存储位置

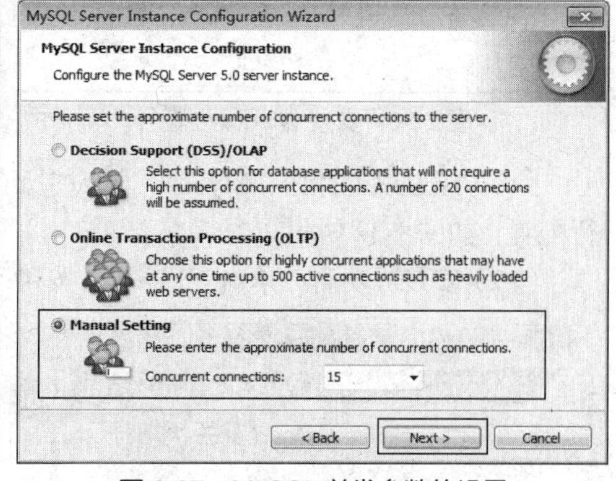


图 2-27　MySQL 并发参数的设置

（14）设置 TCP/IP 参数，勾选“Enable TCP/IP Networking”选项，默认的端口是“3306”，勾选“Enable Strict Mode”选项，单击“Next”按钮，如图 2-28 所示。

（15）配置字符集配置，勾选“Manual Selected Default Character Set / Collation”选项，在“Character Set”下拉列表框中选择“utf8”，单击“Next”按钮，如图 2-29 所示。

（16）将 MySQL 安装为 Windows 服务，勾选“Install As Windows Service”和“Include Bin Directory in Windows PATH”选项，单击“Next”按钮，如图 2-30 所示。

（17）设置管理员密码，管理员账号为 root，在“New root password”文本框中填写密码，并在“Confirm”文本框中再次输入该密码，单击“Next”按钮，如图 2-31 所示。

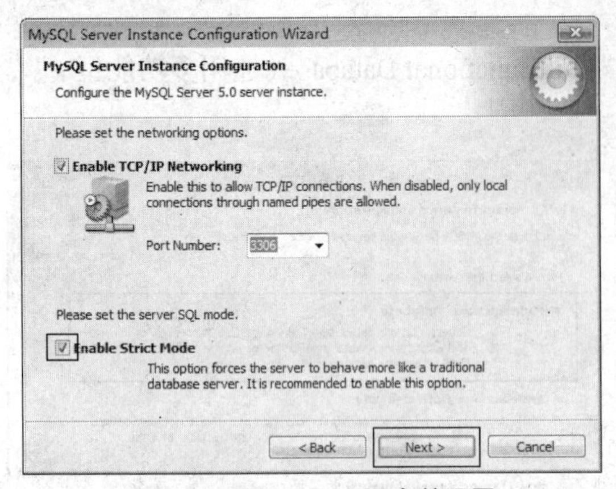


图 2-28　MySQL TCP 参数设置

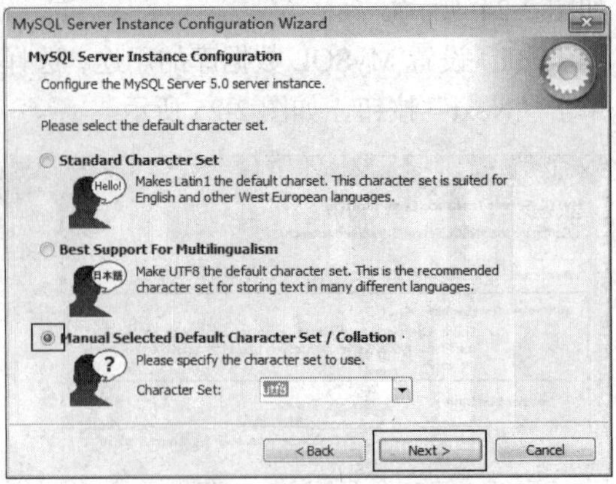

图 2-29　MySQL 的编码设置

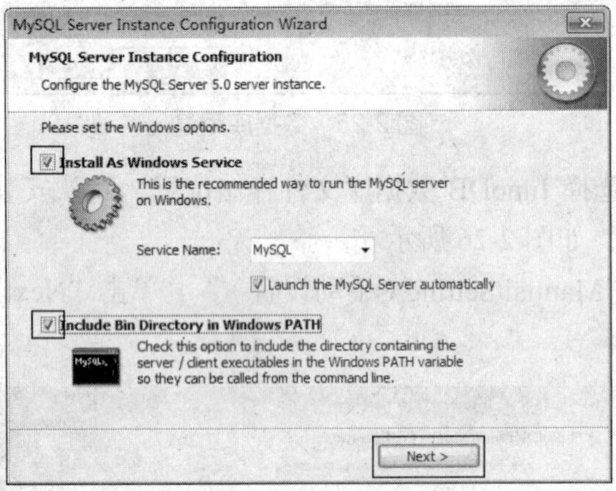

图 2-30　安装 Windows 服务

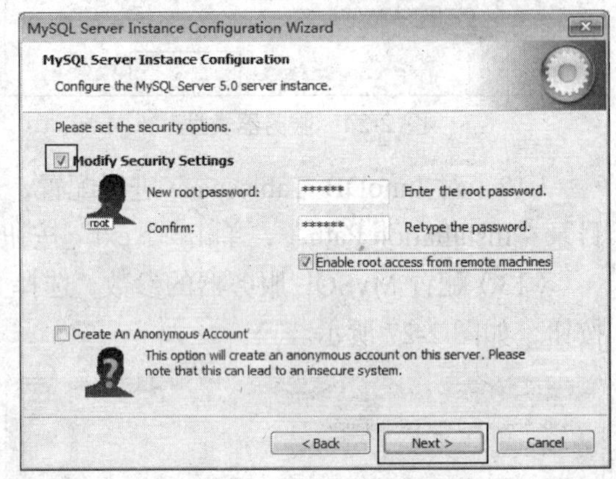


图 2-31　登录密码参数设置

（18）经过以上操作，MySQL 安装的所有配置都已设置完成，单击“Execute”按钮执行配置，如图 2-32 所示。

（19）安装成功后，单击“Finish”按钮，如图 2-33 所示。

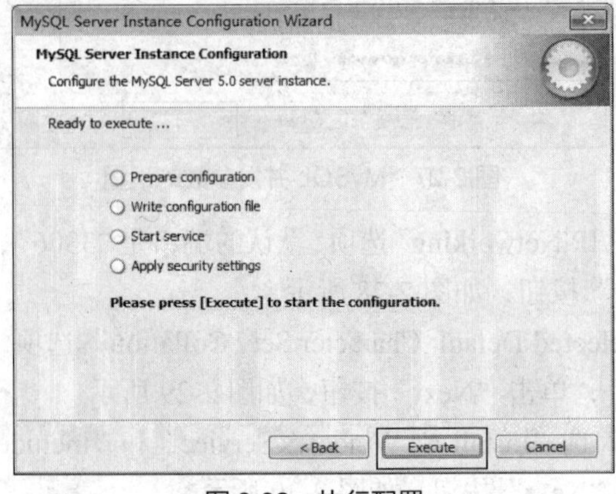


图 2-32　执行配置

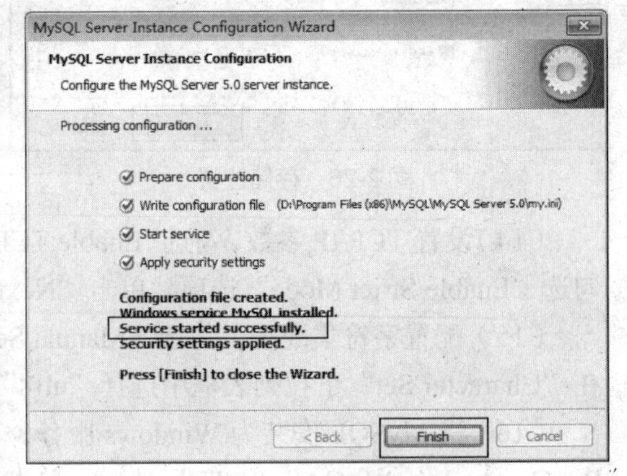


图 2-33　安装成功

### 2. 安装 MySQL 数据库管理工具

使用 Navicat 作为 MySQL 的客户端管理工具，安装步骤如下：

运行 Navicat 8.0.exe 安装程序，设置安装目录后，单击“安装”按钮，如图 2-34 所示。

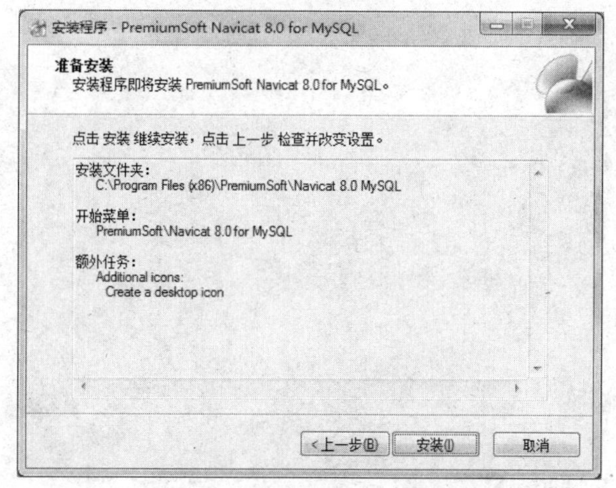


图 2-34　安装 MySQL 可视化工具

### 3. 功能验证测试

（1）打开 Navicat，测试与本地 MySQL 数据库连接，填入主机名、端口号、用户名与密码参数，单击“确定”按钮，如图 2-35 所示。

（2）单击“连接测试”按钮，提示连接成功，如图 2-36 所示。

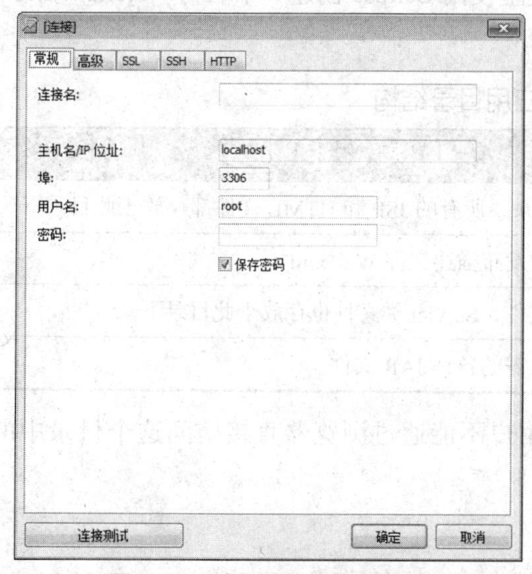

图 2-35　输入数据库连接参数

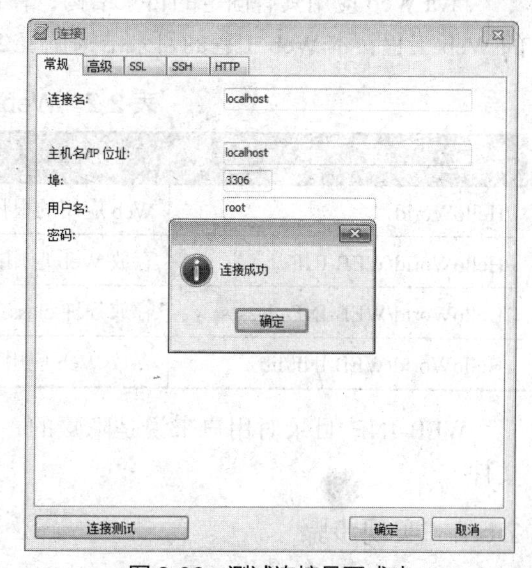


图 2-36　测试连接是否成功

（3）进入到 Navicat 主界面，用户可以对指定的数据库进行操作和管理，如图 2-37 所示。

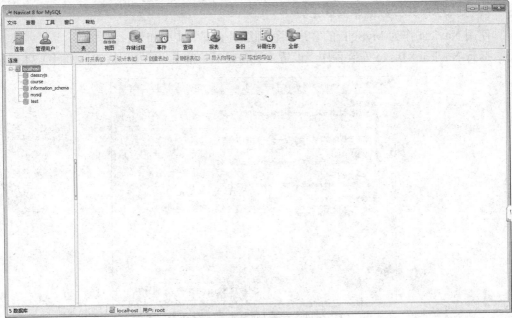


图 2-37　Navicat 主界面

# 任务 2.5　新建 HelloWorld 程序

## 2.5.1　相关知识

Java Web 应用具有固定的目录结构，在这里使用 Eclipse 创建一个名为“HelloWorld”的 Web 工程，本 Web 工程的目录结构见表 2-2。

表 2-2　Web 应用目录结构

| 目　录 | 描　述 |
| --- | --- |
| /HelloWorld | Web 应用的根目录，所有的 JSP 和 HTML 文件都存放于此目录下 |
| /HelloWorld/WEB-INF | 存放 Web 应用的发布描述文件 Web.xml |
| /HelloWorld/WEB-INF/classes | 存放各种 class 文件，Servlet 类文件也存放于此目录下 |
| /HelloWorld/WEB-INF/lib | 存放 Web 应用所需的各种 JAR 文件 |

WEB-INF 目录对用户来说是隐藏的，用户不能通过浏览器直接访问这个目录中的文件。

## 2.5.2　实现步骤

### 1．新建 Hello World 工程

（1）选择 Eclipse 中的“File”→“New”→“Dynamic Web Project”菜单任务，创建一个 Web 工程，如图 2-38 所示。

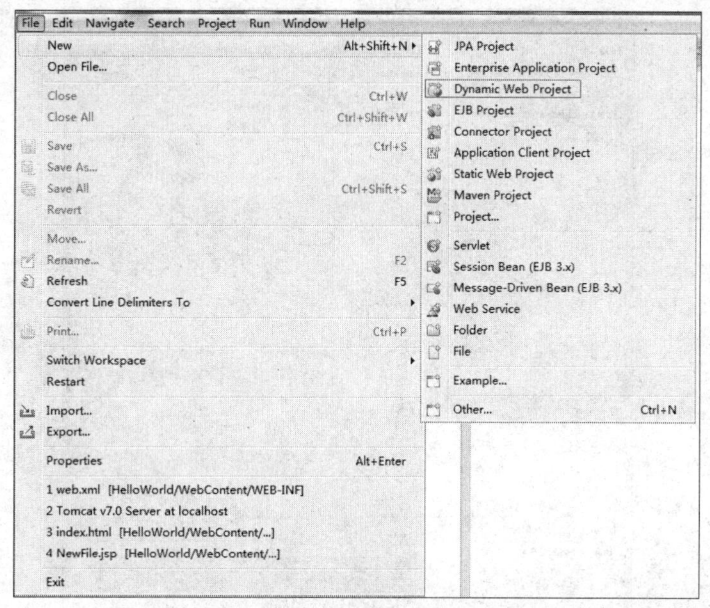


图 2-38　新建 Web 工程

（2）填写项目名称为“HelloWorld”，选择之前配置好的 Tomcat7.0 服务器，单击“Next”按钮，如图 2-39 所示。

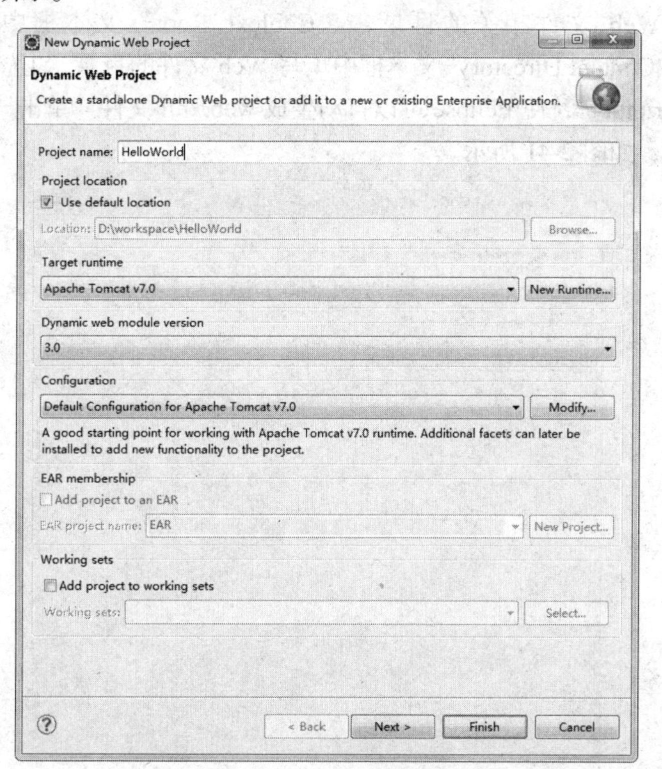


图 2-39　Web 工程配置

（3）设置 Web 项目存放 Java 源代码的目录，默认为 src 目录，单击“Next”按钮，如图 2-40 所示。

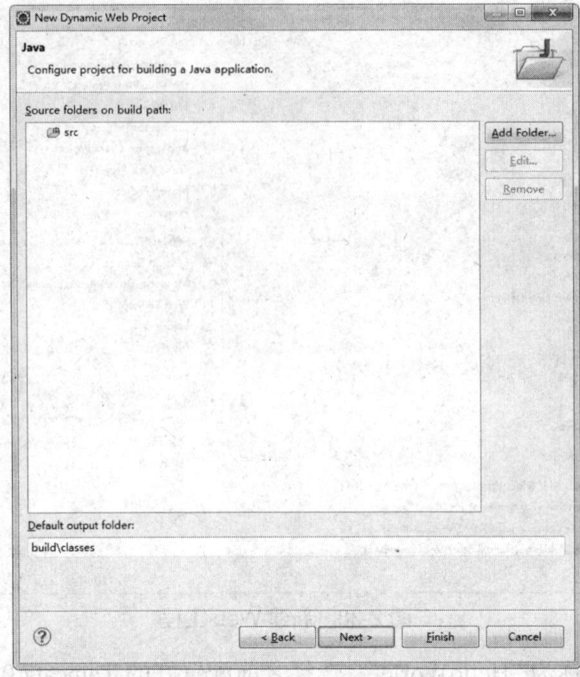


图 2-40　Java 文件目录配置

（4）设置 Web 文件相关的目录，“Context Root”文本框中填写工程的名称“HelloWorld”，“Content Directory”文本框中填写 Web 文件的目录，勾选“Generate web.xml deployment descriptor”，使 Eclipse 可以自动生成 web.xml 文件。单击“Finish”按钮完成项目工程的创建，如图 2-41 所示。

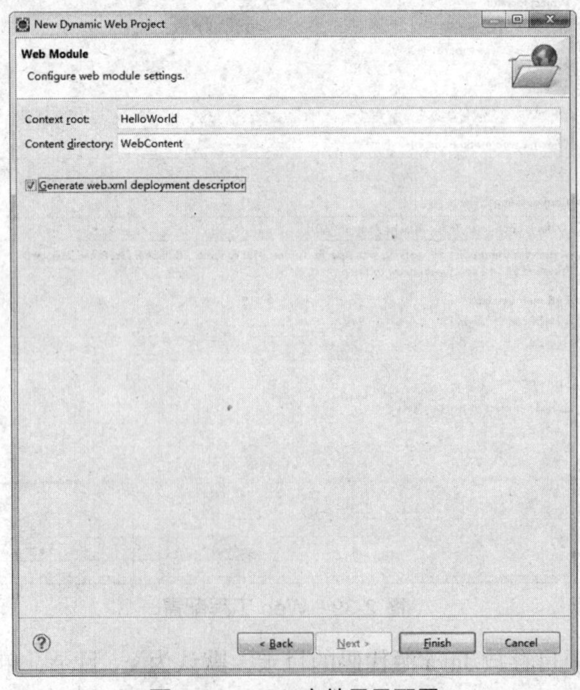


图 2-41　Web 文件目录配置

2. 新建 Web 页面

Web 项目新建完成之后，还需要给项目新建一个可以访问的 Web 页面，命名为 index.html。

（1）新建完成的 Web 项目结构，如图 2-42 所示。

（2）右击项目，在弹出的快捷菜单中选择“New”命令，在 WebContent 目录下创建一个 HTML File 文件，并命名为“index.html”，如图 2-43 所示。

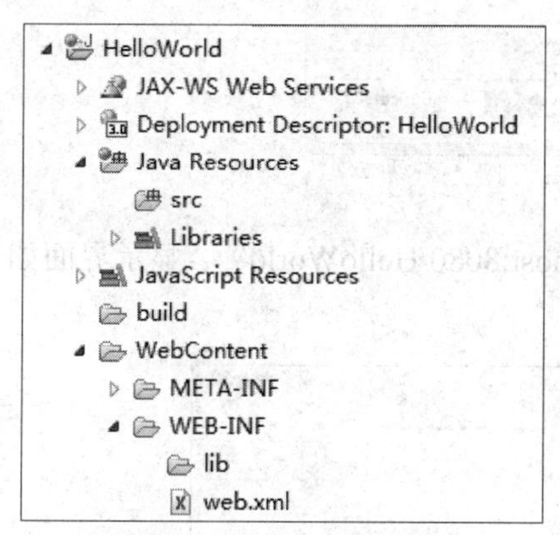


图 2-42　Web 项目的结构

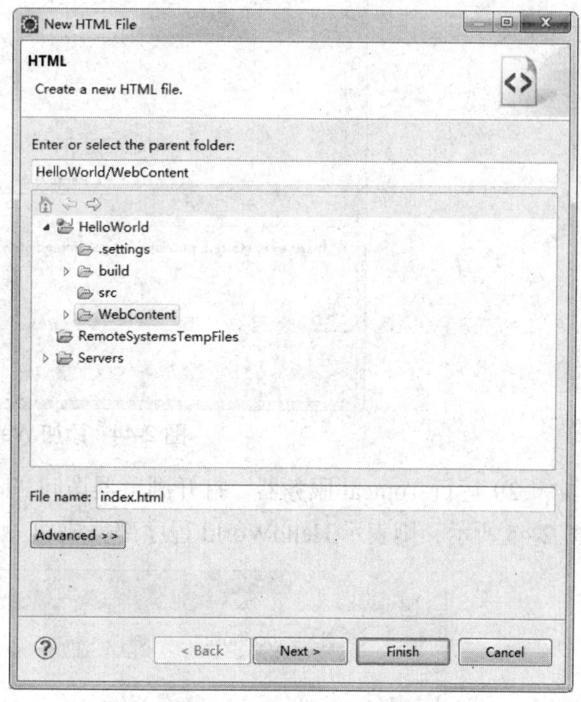


图 2-43　新建 HTML 文件

（3）编辑 index.html 文件，输入如下代码：

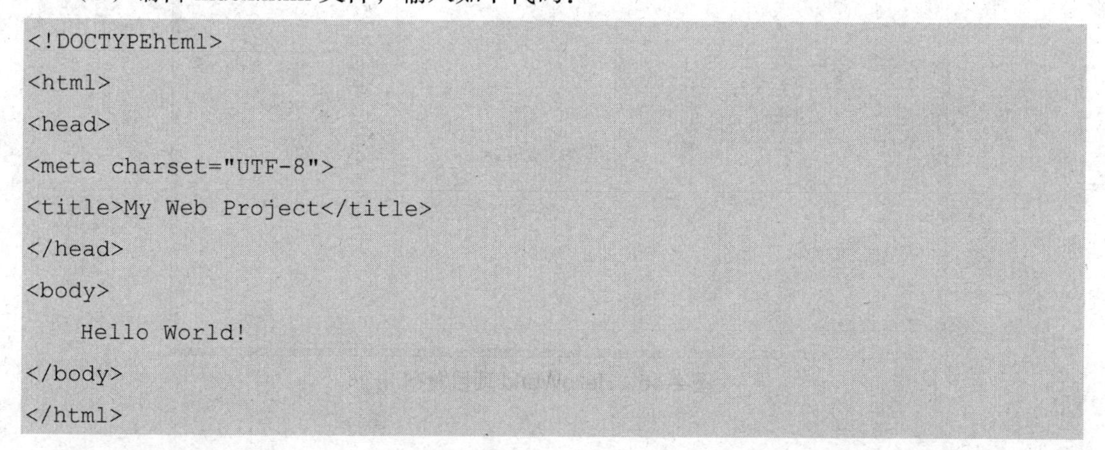

```
<!DOCTYPEhtml>
<html>
<head>
<meta charset="UTF-8">
<title>My Web Project</title>
</head>
<body>
    Hello World!
</body>
</html>
```

3. 功能验证测试

（1）右击 Tomcat，将 HelloWorld 项目添加至 Tomcat，如图 2-44 所示。

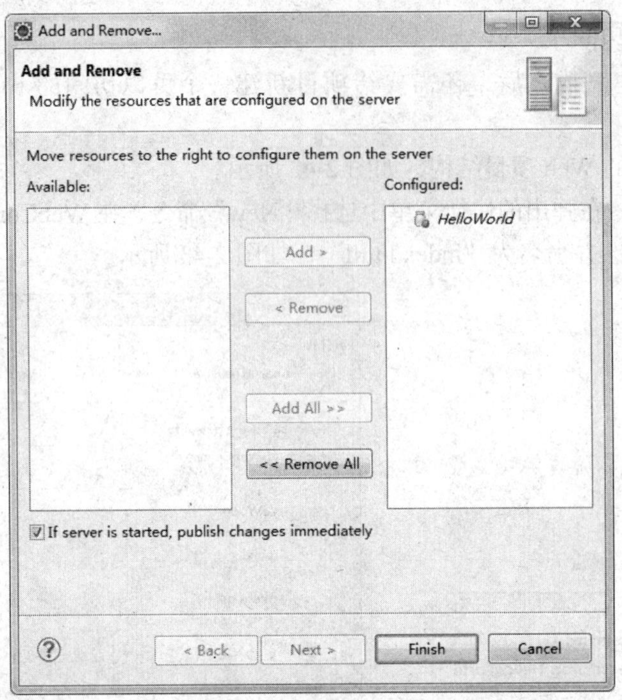


图 2-44　添加 Web 项目

（2）运行 Tomcat 服务器。打开浏览器访问“localhost:8080/HelloWorld”，显示界面如图 2-45 所示，则表示 HelloWorld 程序部署成功。

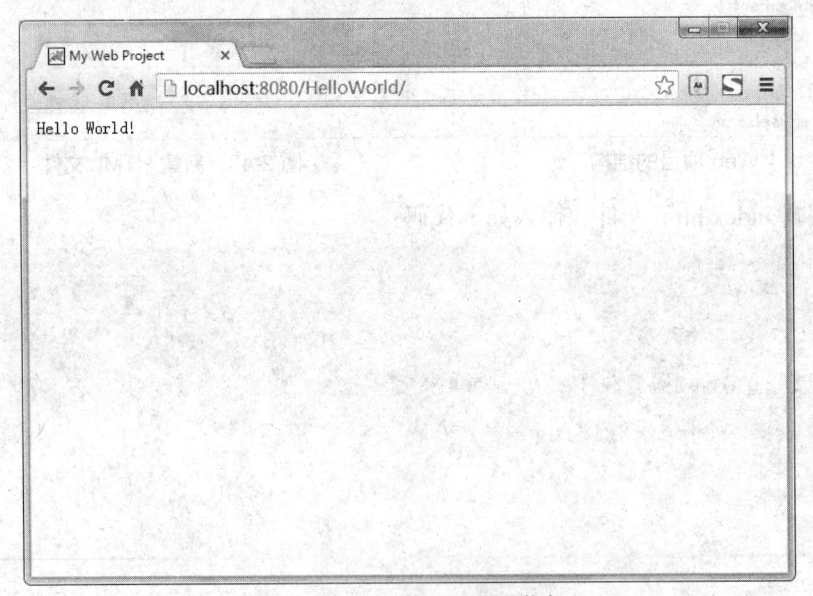


图 2-45　HelloWorld 项目发布

# 项目 3 JavaEE 基础知识

## 单元介绍

本单元读者需要学习 JavaEE 基础知识，掌握 JavaEE 开发的主流技术框架，为后续实现更复杂的功能做准备。

## 学习任务

本单元主要完成以下学习目标：

- 熟悉前端开发框架 Bootstrap3，了解列布局；
- 熟悉 JSTL 标签库的使用；
- 熟悉关系映射框架 Hibernate，掌握 CRUD 的基本操作；
- 熟悉 Spring 框架，了解常用的注解，以及 Spring MVC 的工作原理。

根据以上学习目标，本项目将分解为以下 5 个任务，具体见表 3-1。

表 3-1 任务分解表

| 任务序号 | 任务内容 | 验证方式 |
|---|---|---|
| 任务 3.1 | 了解 Bootstrap3 相关知识 | 实现网盘主界面架构，主界面显示正常 |
| 任务 3.2 | 了解 JSTL 相关知识 | 显示文件类型和文件列表 |
| 任务 3.3 | 了解 Spring 相关知识及实现步骤 | 实现登录界面的控制层，界面显示正常 |
| 任务 3.4 | 了解 Hibernate 相关知识 | 实现登录界面的服务层，界面显示正常 |

## 任务 3.1 了解 Bootstrap3 相关知识

### 3.1.1 相关知识

Bootstrap 是一个用于快速开发 Web 应用程序和网站的前端框架。Bootstrap 提供了 Web 界面开发的基本模块，包括 Grid、Typography、Tables、Forms、Buttons、Responsiveness、Dropdowns、Navigation、Modals、Typehead、Pagination、Carousal、Breadcrumb、Tab、Thumbnails、Headers 等组件。

通过这些组件，开发人员无需掌握太多的 HTML 和 CSS 知识，就可以快速地搭建起一个 Web 界面。需要指出的是，Bootstrap 是一个通用的框架，就像任何其他通用的东西一样，如果需要对该框架进行定制开发，就需要开发人员具备良好的 HTML 和 CSS 基础，才能对这个框架进行深入研究。

在学习这些组件前，首先介绍网盘项目基本的 HTML 结构。

```
<!DOCTYPE html>
<html>
  <head>
    <title>Bootstrap V3 template</title>
    <meta name="viewport" content="width=device-width, initial-scale=1.0">
    <!-- bootstrap css-->
    <link href=" /css/bootstrap.min.css" rel="stylesheet" media="screen">
  </head>
  <body>
    <h1>Hello, world!</h1>
    <!-- jquery -->
    <script src="js/jquery.min.js"></script>
    <!--bootstrap js-->
    <script src="js/bootstrap.min.js"></script>
  </body>
</html>
```

请注意，之所以首先说明结构，是为了简化开发过程，以及项目开发的编写规范。

### 1. Bootstrap 的网格系统

Bootstrap 提供了一套响应式、移动设备优先的流式网格系统，随着屏幕的拉伸，系统会自动分为最多 12 列，如下所示。

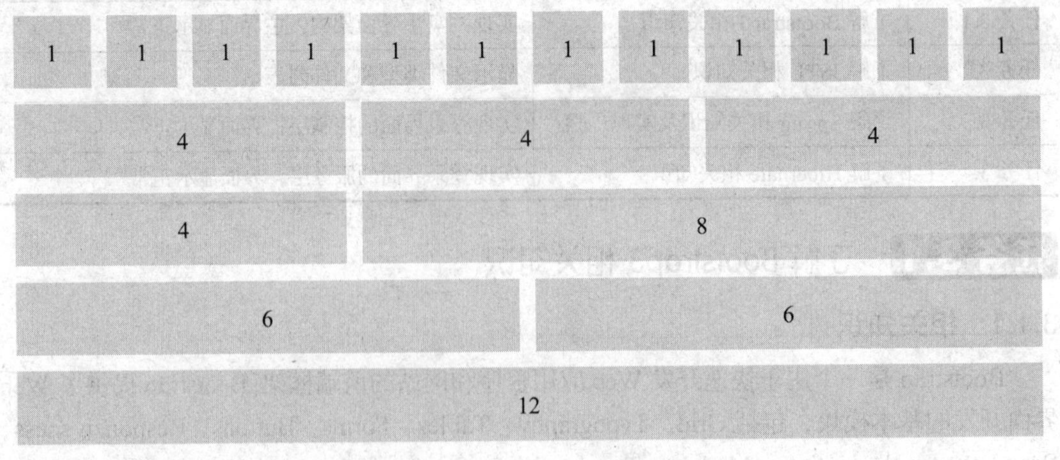


（1）网格选项

表 3-2 总结了 Bootstrap 网格系统如何跨多个设备工作的。

表 3-2 Bootstrap 网格系统

| | 超小设备手机（<768px） | 小型设备平板电脑（≥768px） | 中型设备台式电脑（≥992px） | 大型设备台式电脑（≥1200px） |
|---|---|---|---|---|
| 网格行为 | 一直是水平的 | 以折叠开始，断点以上是水平的 | 以折叠开始，断点以上是水平的 | 以折叠开始，断点以上是水平的 |
| 最大容器宽度 | None(auto) | 750px | 970px | 1170px |
| Class 前缀 | .col-xs- | .col-sm- | .col-md- | .col-lg- |
| 列数量和 | 12 | 12 | 12 | 12 |
| 最大列宽 | Auto | 60px | 78px | 95px |
| 间隙宽度 | 30px（一个列的每边分别为 15px） | 30px（一个列的每边分别为 15px） | 30px（一个列的每边分别为 15px） | 30px（一个列的每边分别为 15px） |
| 可嵌套 | Yes | Yes | Yes | Yes |
| 偏移量 | Yes | Yes | Yes | Yes |
| 列排序 | Yes | Yes | Yes | Yes |

（2）基本的网格结构

下面是 Bootstrap 网格的基本结构：

```
<div class="container">
  <div class="row">
    <div class="col-*-*"></div>
    <div class="col-*-*"></div>
  </div>
  <div class="row">...</div>
</div>
```

2. Bootstrap 的表格

Bootstrap 提供了一个清晰的创建表格的布局。

（1）表格类

Bootstrap 的表格样式见表 3-3。

表 3-3 表格样式

| 类 | 描　述 |
|---|---|
| .table | 为任意<table>添加基本样式（只有横向分隔线） |
| .table-striped | 在<tbody>内添加斑马线形式的条纹（IE8 不支持） |
| .table-bordered | 为所有表格的单元格添加边框 |
| .table-hover | 在<tbody>内的任一行启用鼠标悬停状态 |
| .table-condensed | 让表格更加紧凑 |

（2）<tr>、<th>和<td>类

Bootstrap 表格的提示样式见表 3-4。

表 3-4　Bootstrap 表格的提示样式

| 类 | 描　述 |
| --- | --- |
| .active | 将悬停的颜色应用在行或者单元格上 |
| .success | 表示成功的操作 |
| .info | 表示信息变化的操作 |
| .warning | 表示一个警告的操作 |
| .danger | 表示一个危险的操作 |

（3）基本的表格

如果想要实现一个只带有内边距（padding）和水平分割的基本表，请添加 class.table，如下面实例所示：

```
<table class="table">
   <caption>基本的表格布局</caption>
<thead>
    <tr> <th>名称</th> <th>城市</th> </tr>
</thead>
<tbody>
    <tr> <td>Tanmay</td> <td>Bangalore</td> </tr>
    <tr> <td>Sachin</td> <td>Mumbai</td> </tr>
</tbody>
</table>
```

运行结果如图 3-1 所示。

基本的表格布局

| 名称 | 城市 |
| --- | --- |
| Tanmay | Bangalore |
| Sachin | Mumbai |

图 3-1　表格布局效果图

### 3. Bootstrap 的按钮

（1）按钮类

Bootstrap 提供了一些选项来定义按钮的样式，这些样式也可以应用到<a>、<button>和<input>元素上，如表 3-5 所示。

表 3-5　Bootstrap 的按钮类

| 类 | 描　　述 |
| --- | --- |
| .btn | 为按钮添加基本样式 |
| .btn-default | 默认/标准按钮 |
| .btn-primary | 原始按钮样式（未被操作） |
| .btn-success | 表示成功的动作 |
| .btn-info | 该样式可用于要弹出信息的按钮 |
| .btn-warning | 表示需要谨慎操作的按钮 |
| .btn-danger | 表示一个危险动作的按钮操作 |
| .btn-link | 让按钮看起来像个链接（仍然保留按钮行为） |
| .btn-lg | 制作一个大按钮 |
| .btn-sm | 制作一个小按钮 |
| .btn-xs | 制作一个超小按钮 |
| .btn-block | 块级按钮（拉伸至父元素 100%的宽度） |
| .active | 按钮被单击 |
| .disabled | 禁用按钮 |

下面的实例演示了上面所有的按钮 class：

```
<!-- 标准的按钮 -->
<button type="button" class="btn btn-default">默认按钮</button>
<!-- 提供额外的视觉效果，标识一组按钮中的原始动作 -->
<button type="button" class="btn btn-primary">原始按钮</button>
<!-- 表示一个成功的或积极的动作 -->
<button type="button" class="btn btn-success">成功按钮</button>
<!-- 信息警告消息的上下文按钮 -->
<button type="button" class="btn btn-info">信息按钮</button>
<!-- 表示应谨慎采取的动作 -->
<button type="button" class="btn btn-warning">警告按钮</button>
<!-- 表示一个危险的或潜在的负面动作 -->
<button type="button" class="btn btn-danger">危险按钮</button>
<!-- 并不强调是一个按钮，看起来像一个链接，但同时保持按钮的行为 -->
<button type="button" class="btn btn-link">链接按钮</button>
```

运行结果如图 3-2 所示。

图 3-2　Bootstrap 按钮样式

（2）按钮组

按钮组允许多个按钮被堆叠在同一行上。当需要把按钮对齐在一起时，这就显得非常有用。可以通过 Bootstrap 按钮（Button）插件添加可选的 JavaScript 单选框和复选框样式行为。

表 3-6 总结了 Bootstrap 提供的使用按钮组的一些重要的 class。

表 3-6　Bootstrap 按钮组类

| Class | 描　述 |
| --- | --- |
| .btn-group | 该 class 用于形成基本的按钮组。在.btn-group 中放置一系列带有 class.btn 的按钮 |
| .btn-toolbar | 该 class 有助于把几组<div class="btn-group">结合到一个<div class="btn-toolbar">中，获得更复杂的组件 |
| .btn-group-lg，.btn-group-sm，.btn-group-xs | 这些 class 可应用到整个按钮组的大小调整，而不需要对每个按钮进行大小调整 |
| .btn-group-vertical | 该 class 可以使一组按钮垂直堆叠显示 |

（3）基本的按钮组

下面的实例演示了上面表格中讨论到的 class.btn-group 的使用：

```
<div class="btn-group">
    <button type="button" class="btn btn-default">按钮 1</button>
    <button type="button" class="btn btn-default">按钮 2</button>
    <button type="button" class="btn btn-default">按钮 3</button>
</div>
```

运行结果如图 3-3 所示。

按钮 1　按钮 2　按钮 3

图 3-3　Bootstrap 按钮组

## 3.1.2　实现步骤

本节要实现的网页如图 3-4 所示。

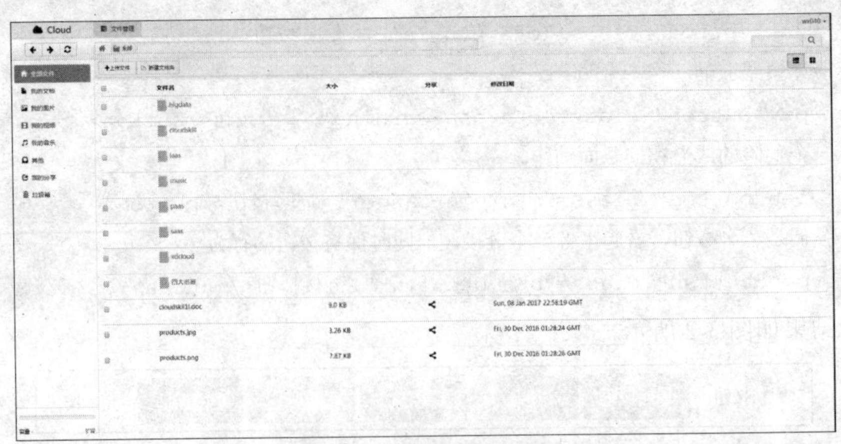


图 3-4　网盘项目主界面

网盘的整个网页由头部、左侧导航、右侧主体三块区域构成。

1. 实现主界面框架

```
<!DOCTYPE html>
<html lang="zh-cn">
<head>
<meta charset="utf-8">
<meta http-equiv="X-UA-Compatible" content="IE=edge">
<meta name="viewport" content="width=device-width, initial-scale=1">
<meta name="description" content="">
<meta name="author" content="">
<title>先电云存储</title>
</head>
<body>
<!--头部区域代码-->
 ...
<div class="main">
<!--左侧导航区域代码-->
...
<!--右侧内容区域代码-->
...
</div>
</body>
</html>
```

2. 引入样式文件及基础 js 文件

在<head></head>标签区域引入相关 CSS 样式文件：

```
<link href="assets/stylesheets/bootstrap.min.css" rel="stylesheet"type="text/
css" />
<link href="assets/stylesheets/style.css" rel="stylesheet" type="text/css" />
<link href="assets/stylesheets/disk.css" rel="stylesheet" type="text/css" />
<link href="assets/stylesheets/font-awesome.css" rel="stylesheet" type="text/
css" />
<link href="assets/stylesheets/zTreeStyle.css" rel="stylesheet" type="text/
css" />
```

在<body></body>标签区域引入相关 javascript 文件：

```
<script src="assets/javascripts/jquery.min.js"></script>
<script src="assets/javascripts/bootstrap.min.js"></script>
<script src="assets/javascripts/fineuploader.js"></script>
```

```
<script src="assets/javascripts/checkbox.js"></script>
<script src="assets/javascripts/common.js"></script>
```

3. 实现头部区域

引用好基础样式及 js 脚本文件后开始编写网页头部代码。头部区域由 topbar 部分和 frame-header 部分组成，topbar 区域的 html 代码：

```
<div class="topbar">
 <div class="content">
 <div class="top_left">
 <a href="javascript:void(0)" class="topbar_menu title" draggable="false">
 <i class="fa fa-cloud"></i>Cloud
 </a>
 <a class="topbar_menu this" target="_self" draggable="false">
 <i class="font-icon menu-explorer"></i>文件管理
 </a>
 </div>
 <div class="top_right">
 <div class="menu_group">
 <a href="#" id="topbar_user" data-toggle="dropdown"
    class="topbar_menu" draggable="false">
 <i class="font-icon icon-user"></i>管理员 <b class="caret"></b></a>
 <ul class="dropdown-menu menu-topbar_user pull-right animated menuShow"
role="menu" aria-labelledby="topbar_user">
 <li>
 <a href="#" draggable="false">
 <i class="font-icon fa fa-sign-out"></i>个人信息</a>
 </li>
 <li>
 <a href="#" draggable="false">
 <i class="font-icon fa fa-sign-out"></i>退出</a></li></ul>
 </div>
 </div>
 <div style="clear: both"></div>
 </div>
 </div>
```

frame-header 区域代码：

```
<div class="frame-header">
<div class="header-content">
<div class="header-left">
```

```
<div class="btn-group btn-group-sm">
<button onclick="javascript:history.go(-1);" class="btn btn-default"
id="history_back" title="后退" type="button">
<i class="font-icon fa fa-arrow-left"></i>
</button>
<button onclick="javascript:history.go(1);" class="btn btn-default"
   id="history_Next " title="前进" type="button">
<i class="font-icon fa fa-arrow-right"></i></button>
<button onclick="javascript:location.reload();" class="btn btn-default"
id="refresh" title="强制刷新"  type="button">
<i class="font-icon fa fa-refresh"></i>
</button>
</div>
</div>
<div class="header-middle">
<a href="#">
<button class="btn btn-default" id="home" title="我的文档">
<i class="font-icon fa fa-home"></i></button>
</a>
<div id="yarnball">
<ul class="yarnball">
<li class="yarnlet first">
<a title="" href="#">
<span class="left-yarn"></span>
<span class="address_ico groupSelf"></span>
<span class="title_name">全部</span>
</a></li></ul></div>
<div id="yarnball_input" style="display: none;">
<input type="text" name="path" value="" class="path" id="path">
</div></div>
<div class="header-right">
<input type="text" name="seach" id="navbarInput-01">
<a class="btn btn-default" id="searchfiles" title="搜索" type="button">
<i class="font-icon fa fa-search"></i>
</a></div></div></div>
```

#### 4. 实现导航区域

编写好头部区域后进入侧边导航栏的编写，首先分析左侧导航区域内容，可以看到导航区域主要由 8 个导航组成。

```
<div class="main-left" style="z-index: 1;">
 <div class="main-left" style="z-index: 1;">
 <div class="list-group category" style="margin: 10px;">
 <div id="ahref">
 <a class="list-group-item active" id="r_active">
  <span class="glyphicon glyphicon-home left-icon"></span>
     全部文件 </a>
 <a class="list-group-item " id="type">
  <span class="glyphicon glyphicon-file left-icon"></span>
    我的文档</a>
 <a class="list-group-item ">
 <span class="glyphicon glyphicon-picture left-icon"></span>
    我的图片</a>
 <a class="list-group-item " id="type">
 <span class="glyphicon glyphicon-film left-icon"></span>
   我的视频</a>
 <a class="list-group-item " id="type">
 <span class="glyphicon glyphicon-music left-icon"></span>
   我的音乐</a>
 <a class="list-group-item" id="type">
 <span class="glyphicon glyphicon-inbox left-icon"></span>
   其他</a>
 <a class="list-group-item" id="type">
 <span class="glyphicon glyphicon-share left-icon"></span>
   我的分享</a>
 <a class="list-group-item" >
 <span class="glyphicon glyphicon-trash left-icon"></span>
   垃圾箱</a></div></div>
 <div class="main-left-use">
 <div class="progress progress-u progress-xs">
 <div class="progress-bar progress-bar-blue" id="progress-bar"
 role="progressbar" aria-valuemin="0" aria-valuemax="100"></div>
 </div>
 <h3 class="heading-xs">容量:
 <span id="totle"></span>
 <span class="capacity pull-right">
 <a href="javascript:void(0)">扩容</a>
 </span></h3></div></div></div>
```

### 5. 实现文件主体内容区域

文件主体内容区域代码主要分为两部分，一部分是按钮组，另一部分是文件列表。之前已经介绍过 Bootstrap 相关按钮组件及表格组件，下面来使用这些相关组件。

文件内容区域主结构如下所示：

```
<div class="main-right">
   <!--按钮组-->
      <div class="col-md-12" style="padding: 0"></div>
   <!--缩略图按钮-->
      <div class="col-md-12" style="padding: 0"></div>
   <!--文件列表-->
     <div class="col-md-12" style="padding: 0"></div>
   </div>
```

按钮组包括上传、新建文件夹、删除、下载、重命名、复制、移动、数据分析等按钮，是网盘操作的主要入口。按钮组的 html 代码如下：

```
<div class="tools">
 <div class="tools-left">
 <div class="btn-group btn-group-sm kod_path_tool fl-left">
 <div id="result-uploader" class="right upload-filemain fl pull-left"></div>
 <button class="btn btn-default" type="button" id="newdir">
 <i class="font-icon fa  fa-folder-open-o"></i>新建文件夹
 </button>
 <button class="btn btn-default" style="display: none;"type="button"id="delete">
 <i class="font-icon glyphicon glyphicon-trash" style="color: #888"></i>删除
 </button>
 <a class="btn btn-default" type="button" style="display: none;" id="download">
 <i class="font-icon glyphicon glyphicon-download-alt" style="color: #888"></i>
 下载
 </a>
 <button  class="btn  btn-default"  style="display:  none;"  type="button"
id="rename">
 <i class="font-icon glyphicon glyphicon-pencil" style="color: #888"></i>
 重命名
 </button>
 <a class='btn btn-default' style="display: none;" id="download">
 <span lass="glyphicon glyphicon-download-alt"><span>下载文件</a>
 <a class='btn btn-default' style="display: none;" id="copy">
 <span class="glyphicon glyphicon-file"></span>复制</a>
 <a class='btn btn-default' style="display: none;" id="move">
```

```
<span class="glyphicon glyphicon-transfer"></span>移动</a>
<a class='btn btn-default' style="display: none;" id="rename">
<span class="glyphicon glyphicon-pencil"></span>重命名</a>
<a class='btn btn-default' style="display: none" id="Dataanalysis"></a>
<ul class="pull-left">
<li class="dropdown">
<a class="dropdown-toggle btn btn-default" data-toggle="dropdown"
  id="selectColumn" style="display: none" href="javascript:void(0);">
 数据分析</a>
<ul class="dropdown-menu cloudanys" style="margin-top: 4px !important;">
<li><a href="javascript:void(0);">词云分析</a></li>
<li class="divider"></li>
<li><a href="javascript:void(0);">柱状分析</a></li>
<li class="divider"></li>
<li><a href="javascript:void(0);">气泡分析</a></li>
</ul></li></ul></div>
<div class="clearfix"></div>
</div></div>
```

文件列表区域主要使用 Bootstrap 表格布局，文件列表的 html 代码如下：

```
<table class="table mb-0">
<thead>
<tr><th class="table-checkbox"style="position: relative; left: 13px;">
<labelclass="checkbox checkbox-position" for="checkbox1">
<span class="icons icon-span">
<span class="first-icon fui-checkbox-unchecked"></span>
<span class="second-icon fui-checkbox-checked"></span>
</span> <input name="chkAll" type="checkbox" id="operAll"
value="checkbox" class="main-check" /></label></th>
<th class="mainfile-name">文件名</th>
<th class="hide table-fileposition mainfile-position">文件位置</th>
<th class="mainfile-size">大小</th>
<th class="mainchange-date">修改日期</th>
</tr>
</thead>
<tbody id="tab">
<tr><td><label class="checkbox table-checkboxposition"for="checkbox1">
<span class="icons main-icons">
<span class="first-icon fui-checkbox-unchecked"></span>
<span class="second-icon fui-checkbox-checked"></span>
</span> <input type="checkbox" name='check' class="main-tabinput">
</label></td>
```

```
<td><span style="display: block">cloud.doc</span>
<div class="edit-name" style="display: none;">
<input class="box" type="text" value="cloud.doc">
<a class="sure" href="javascript:void(0);">
<span class="glyphicon glyphicon-ok"></span>
</a> <a class="cancel ml-10" href="javascript:void(0);">
<span class="glyphicon glyphicon-remove"></span>
</a></div></td><td class="hide table-fileposition table-path">cloudskill/</td>
<td>23 Bytes</td>
<td>2017-01-11</td>
</tr>
</tbody>
</table>
```

最后，在控制层之中插入一段跳转的代码，如下所示：

```
@RequestMapping("/getlogin")
public ModelAndView login(HttpServletRequest request) {
    ModelAndView view = new ModelAndView();
    view.setViewName("main");
    return view;
}
```

### 6. 功能验证测试

将项目部署在 Tomcat 中并发布，在浏览器之中输入 URL 地址，查看网页效果，最终效果如图 3-5 所示。

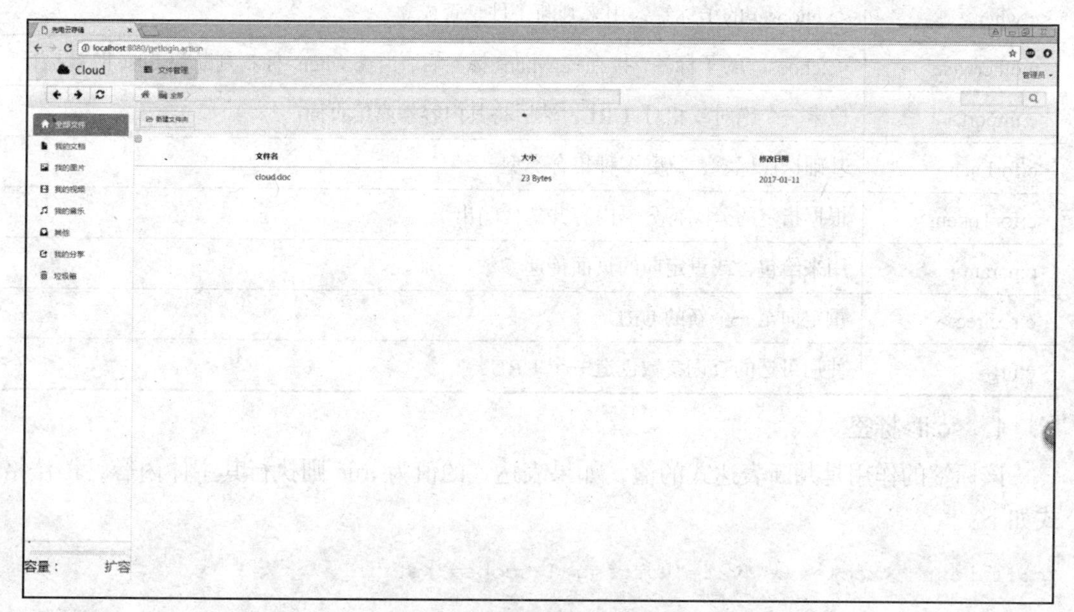


图 3-5　网盘主界面运行效果

## 任务 3.2 了解 JSTL 相关知识

### 3.2.1 相关知识

JSP 标准标签库（JSTL）是一个 JSP 标签集合，它封装了 JSP 应用的通用核心功能。

JSTL 支持通用的、结构化的任务，比如迭代、条件判断、XML 文档操作、国际化标签、SQL 标签。除这些之外，它还提供了一个框架来使用集成 JSTL 的自定义标签。

根据 JSTL 标签所提供的功能，可以将其分为 5 个类别：核心标签、格式化标签、SQL 标签、XML 标签和 JSTL 函数。

核心标签是最常用的 JSTL 标签。引用核心标签库的语法如下：

```
<%@ taglib prefix="c"  uri="http://java.sun.com/jsp/jstl/core" %>
```

核心标签的使用见表 3-7。

表 3-7　JSTL 标签说明

| 标　签 | 描　述 |
| --- | --- |
| <c:out> | 用于在 JSP 中显示数据，就像<%= ... > |
| <c:set> | 用于保存数据 |
| <c:remove> | 用于删除数据 |
| <c:catch> | 用来处理产生错误的异常状况，并且将错误信息存储起来 |
| <c:if> | 与一般程序中用的 if 一样 |
| <c:choose> | 本身只当作<c:when>和<c:otherwise>的父标签 |
| <c:when> | <c:choose>的子标签，用来判断条件是否成立 |
| <c:otherwise> | <c:choose>的子标签，接在<c:when>标签后，当<c:when>标签判断为 false 时被执行 |
| <c:import> | 检索一个绝对或相对 URL，然后将其内容暴露给页面 |
| <c:forEach> | 基础迭代标签，接受多种集合类型 |
| <c:forTokens> | 根据指定的分隔符分隔内容并迭代输出 |
| <c:param> | 用来给包含或重定向的页面传递参数 |
| <c:redirect> | 重定向至一个新的 URL |
| <c:url> | 使用可选的查询参数创造一个 URL |

#### 1. <c:if>标签

该标签的作用是判断表达式的值，如果表达式的值为 true 则执行其主体内容，语法格式如下：

```
<c:if test="<boolean>" var="<string>" scope="<string>">
   ...
</c:if>
```

该标签的属性见表 3-8。

表 3-8 <c:if>标签的属性

| 属 性 | 描 述 | 是否必要 | 默认值 |
| --- | --- | --- | --- |
| test | 条件 | 是 | 无 |
| var | 用于存储条件结果的变量 | 否 | 无 |
| scope | var 属性的作用域 | 否 | page |

### 2. <c:forEach>标签

该标签封装了 for、while、do-while 循环，语法格式如下：

```
<c:forEach
   items="<object>"
   begin="<int>"
   end="<int>"
   step="<int>"
   var="<string>"
   varStatus="<string>">
   ...
```

该标签的属性见表 3-9。

表 3-9 <c:forEach>标签的属性

| 属性 | 描 述 | 是否必要 | 默认值 |
| --- | --- | --- | --- |
| items | 要被循环的信息 | 否 | 无 |
| begin | 开始的元素（0=第一个元素，1=第二个元素） | 否 | 0 |
| end | 最后一个元素（0=第一个元素，1=第二个元素） | 否 | Last element |
| step | 每一次迭代的步长 | 否 | 1 |
| var | 代表当前条目的变量名称 | 否 | 无 |
| varStatus | 代表循环状态的变量名称 | 否 | 无 |

### 3. <c:choose>标签

<c:choose>标签与 Java switch 语句的功能一样，用于在众多选项中做出选择。

Java switch 语句中有 case 语句，而<c:choose>对应有<c:when>标签，switch 语句中有 default 语句，而<c:choose>标签对应有<c:otherwise>标签。该标签的语法格式如下：

```
<c:choose>
   <c:when test="<boolean>"/>
      ...
   </c:when>
```

```
<c:when test="<boolean>"/>
    ...
</c:when>
...
...
<c:otherwise>
    ...
</c:otherwise></c:choose>
```

<c:choose>标签和<c:otherwise>标签没有属性。<c:when>标签只有一个属性，见表 3-10。

表 3-10　<c:when>标签的属性

| 属性 | 描　述 | 是否必要 | 默认值 |
|---|---|---|---|
| test | 条件 | 是 | 无 |

### 3.2.2　实现步骤

#### 1. 实现文件类型显示

在头部区域有一块内容是根据当前不同的文件类型，显示不同的效果，这里用到了“<c:if>”方法，根据后台传过来的 type 类型来显示不同的 html 代码。

```
<div id="yarnball" title="" style="display: block;">
                <ul class="yarnball">
                <li class="yarnlet first"><a title="" href="home.action">
                <span class="left-yarn"></span>
                <span class="address_ico groupSelf"></span>
                <span class="title_name">全部</span>
                </a></li>
                <c:if test="${type == 2}">
                <li class="yarnlet "><a title="" href="category.action?type=2">
                <span class="title_name">文档</span></a></li>
                </c:if>
                <c:if test="${type == 1}">
                <li class="yarnlet "><a title="" href="category.action?type=1">
                <span class="title_name">图片</span></a></li>
                </c:if>
                <c:if test="${type == 3}">
                <li class="yarnlet "><a title="" href="category.action?type=3" >
                <span class="title_name">视频</span></a></li>
                </c:if>
                <c:if test="${type == 4}">
```

```
        <li class="yarnlet "><a title="" href="category.action?type=4">
        <span class="title_name">音乐</span></a></li>
        </c:if>
        <c:if test="${type == 5}">
        <li class="yarnlet "><a title="" href="category.action?type=5">
        <span class="title_name">其他</span></a></li>
        </c:if>
        <c:if test="${type == 6}">
        <li class="yarnlet "><a title="" href="shareFile.action">
        <span class="title_name">我的分享</span></a></li>
        </c:if>
        <c:if test="${type == 7}">
        <li class="yarnlet "><a title="" href="garbage.action">
        <span class="title_name">垃圾箱</span></a></li>
        </c:if>
            </ul>
            </div>
```

代码运行的效果如图 3-6 所示。

图 3-6　<c:if>标签运行效果

### 2. 实现文件列表显示

在网盘中文件列表通过<c:forEach>及<c:when>方法循环获取个人网盘中的内容，新建一个名为 jstl 的 jsp 页面，在页面的 body 部分插入如下代码：

```
<tbody id="tab">
    <c:forEach var="fb" items="${list}">
        <tr>
            <td><label class="checkbox table-checkboxposition" for="checkbox1">
            <span class="icons main-icons">
            <span class="first-icon fui-checkbox-unchecked"></span>
            <span class="second-icon fui-checkbox-checked"></span>
             </span>
  <input type="checkbox" name='check'class="main-tabinput"onclick="show()">
             </label>
        </td>
```

```
<td><span style="display: block">
<c:choose>
    <c:when test="${fb.isdirectory == true}">
    <a href="home.action?path=${fb.path}">
  </c:when>
  <c:otherwise>
    <a href="javascript:void(0);" onclick="showread('${fb.path}');">
  </c:otherwise>
</c:choose>
<c:if test="${fb.isdirectory == true}">
    <img src="assets/images/0.png" class="objimg">
<input name="objimg" type="text" style="display: none" value="${fb.name}">
</c:if> </a> ${fb.name}</span>
<div class="edit-name" style="display: none;">
    <input class="box" type="text" value="${fb.name}">
    <a class="sure" href="javascript:void(0);" onclick="sure()">
      <span class="glyphicon glyphicon-ok"></span></a>
    <a class="cancel ml-10" onclick="cancel()" href="javascript:void(0);">
      <span class="glyphicon glyphicon-remove"></span>
    </a>
  </div></td>
<td class="hide table-fileposition table-path">${fb.path}</td>
    <td>${fb.length}</td>
  <td id="share_td"><c:if test="${fb.length != null}">
        <a type="button"><img src="assets/images/share2.png"/></a>
     </c:if>
  </td>
    <td>${fb.lastmodified }</td>
    </tr>
    </c:forEach>
    </tbody>
```

在控制层之中添加代码，返回 list 数组在前台遍历。

```
@RequestMapping("/test")
    public ModelAndView getcs(HttpServletRequest request,
            HttpServletResponse response) {
        ModelAndView mv = new ModelAndView();
        mv.setViewName("/jstl");
        List<Map<String,String>> list =new ArrayList<Map<String,String>>();
```

```
    for(int i=0;i<5;i++){
        Map<String,String> map=new HashMap<String, String>();
        map.put("name","name"+i);
        map.put("size", "size"+i);
        map.put("date", "date"+i);
        list.add(map);
    }
    mv.addObject("list", list);
    return mv;
}
```

3．功能验证测试

将项目部署 Tomcat 中并发布，在浏览器中输入对应的 URL 地址，显示效果如图 3-7 所示。

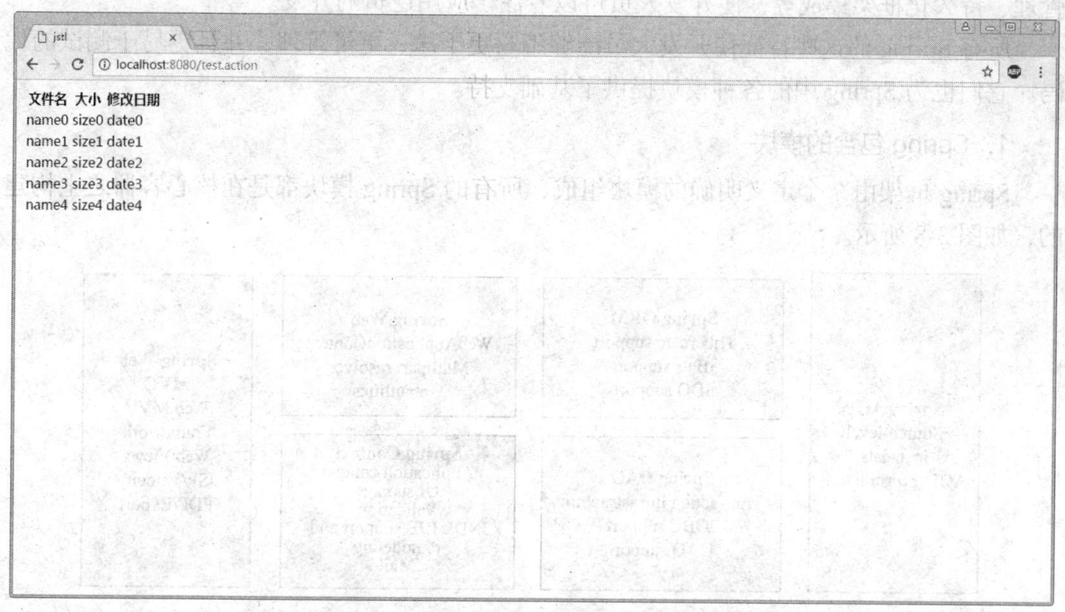


图 3-7 <c:forEach>和<c:when>标签运行效果

## 任务 3.3 了解 Spring 相关知识及实现步骤

### 3.3.1 Spring 相关知识

Spring 是一个开源框架，它由 Rod Johnson 创建。它是为了解决企业应用开发的复杂性而创建的。Spring 使用基本的 JavaBean 完成以前只可能由 EJB 完成的功能。然而，Spring 的用途不限于服务器端的开发。从简单性、可测试性和松耦合的角度而言，任何 Java 应用都可以从 Spring 中受益。

实际上，Spring 是一个轻量级的控制反转（IoC）和面向切面（AOP）的容器框架。Spring 框架的特点如下。

● 轻量：从大小与开销两方面而言，Spring 都是轻量的。完整的 Spring 框架可以在一个大小只有 1MB 多的 JAR 文件中发布，并且 Spring 所需的处理开销也是微不足道的。此外，Spring 是非侵入式的，典型地，Spring 应用中的对象不依赖于 Spring 的特定类。

● 控制反转：Spring 通过一种称作控制反转（IoC）的技术促进了松耦合。当应用了 IoC，一个对象依赖的其他对象会通过被动的方式传递进来，而不是这个对象自己创建或者查找依赖对象。

● 面向切面：Spring 提供了面向切面编程的丰富支持，允许通过分离应用的业务逻辑与系统级服务进行内聚性的开发。

● 容器：Spring 包含并管理应用对象的配置和生命周期，在这个意义上它是一种容器，开发人员可以配置每个 Bean 如何被创建。然而，Spring 不应该被混同于传统的重量级的 EJB 容器，它们经常是庞大与笨重的，难以使用。

● 框架：Spring 可以将简单的组件配置组合成为复杂的应用。在 Spring 中，应用对象被声明式地组合，典型的是在一个 XML 文件里。Spring 也提供了很多基础功能，如事务管理、持久化框架集成等，使开发人员可以专注于应用逻辑的开发。

所有 Spring 的这些特征使开发人员能够编写更干净、更可管理，并且更易于测试的代码。它们也为 Spring 中的各种模块提供了基础支持。

### 1. Spring 包含的模块

Spring 框架由 7 个定义明确的模块组成，所有的 Spring 模块都是在核心容器之上构建的，如图 3-8 所示。

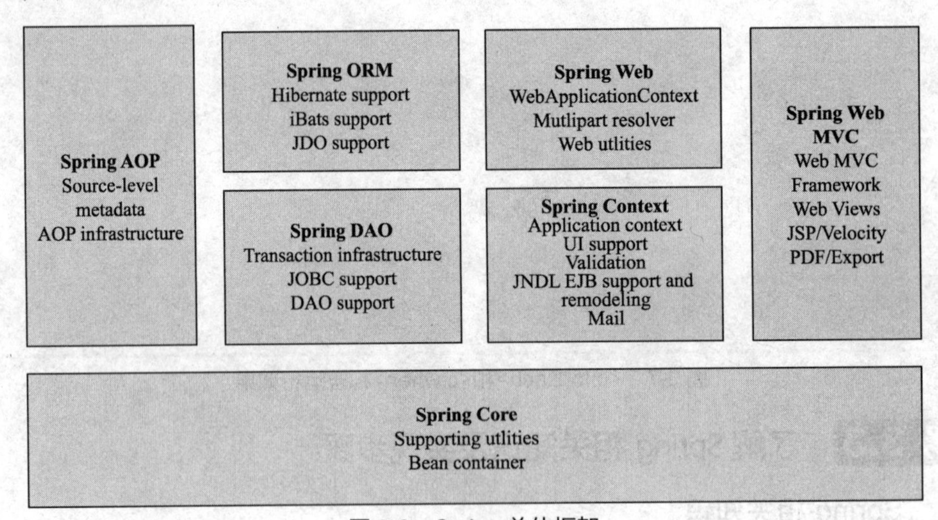


图 3-8 Spring 总体框架

### 2. 核心容器

这是 Spring 框架最基础的部分，它提供了依赖注入（Dependency Injection）特征来实现容器对 Bean 的管理。这里最基本的概念是 BeanFactory，它是任何 Spring 应用的核心。BeanFactory 是工厂模式的一个实现，它使用 IoC 将应用配置和依赖说明从实际的应用代码中分离出来。

### 3. 应用上下文（Context）模块

核心模块的 BeanFactory 使 Spring 成为一个容器，而上下文模块使它成为一个框架。这个模块扩展了 BeanFactory 的概念，增加了对国际化消息、事件传播，以及验证的支持。

另外，这个模块提供了许多企业服务，例如，电子邮件、JNDI 访问、EJB 集成、远程，以及时序调度（scheduling）服务，也包括了对模版框架如 Velocity 和 FreeMarker 集成的支持。

### 4. Spring 的 AOP 模块

Spring 在它的 AOP 模块中提供了对面向切面编程的丰富支持。这个模块是在 Spring 应用中实现切面编程的基础。为了确保 Spring 与其他 AOP 框架的互用性，Spring 的 AOP 支持基于 AOP 联盟定义的 API。AOP 联盟是一个开源项目，它的目标是通过定义一组共同的接口和组件来促进 AOP 的使用，以及不同的 AOP 实现之间的互用性。通过访问站点 http://aopalliance.sourceforge.net，可以找到关于 AOP 联盟的更多内容。

Spring 的 AOP 模块也将元数据编程引入了 Spring。使用 Spring 的元数据支持，可以为源代码增加注释，指示 Spring 在何处，以及如何应用切面函数。

### 5. JDBC 抽象和 DAO 模块

使用 JDBC 经常导致大量的重复代码，取得连接、创建语句、处理结果集，然后关闭连接。由于 Spring 的 JDBC 和 DAO 模块抽取了这些重复代码，因此可以保持数据库访问代码的干净简洁，并且可以防止因关闭数据库资源失败而引起的问题。

### 6. 对象/关系映射集成模块

对那些更喜欢使用对象/关系映射工具而不是直接使用 JDBC 的人，Spring 提供了 ORM 模块。Spring 并不试图实现它自己的 ORM 解决方案，而是为几种流行的 ORM 框架提供了集成方案，包括 Hibernate、JDO 和 iBATIS SQL 映射。Spring 的事务管理支持这些 ORM 框架中的每一个也包括 JDBC。

### 7. Spring 的 Web 模块

Web 上下文模块建立于应用上下文模块之上，提供了一个适合于 Web 应用的上下文。另外，这个模块还提供了一些面向服务支持。例如：实现文件上传的 multipart 请求，它也提供了 Spring 和其他 Web 框架的集成，比如，Struts、WebWork。

### 8. Spring 的 MVC 框架

Spring 为构建 Web 应用提供了一个功能全面的 MVC 框架。虽然 Spring 可以很容易地与其他 MVC 框架集成，例如，Struts，但 Spring 的 MVC 框架使用 IoC 对控制逻辑和业务对象提供了完全的分离。

### 9. Spring Web MVC 架构

Spring Web MVC 框架是一个基于请求驱动的 Web 框架，并且也使用了前端控制器模式进行设计，再根据请求映射规则分发给相应的页面控制器（动作/处理器）进行处理。

Spring Web MVC 处理请求的流程如图 3-9 所示。

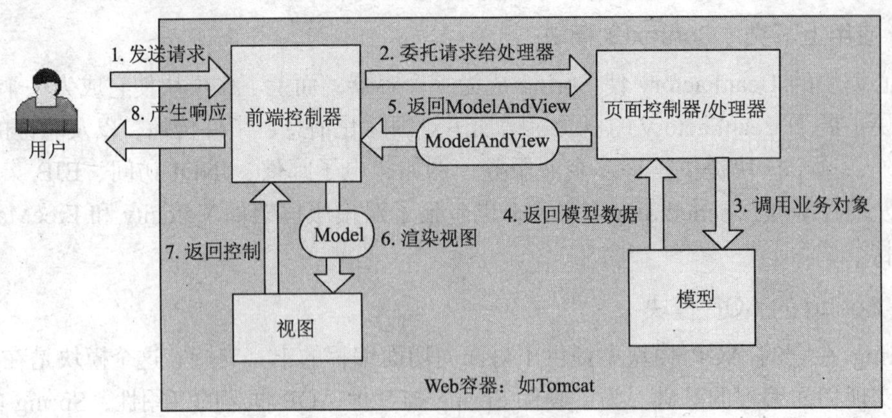


图 3-9　MVC 的工作流程

具体执行步骤如下。

（1）步骤 1、2：用户发送请求到前端控制器，前端控制器根据请求信息（如 URL）决定选择哪一个页面控制器进行处理并把请求委托给它，即以前的控制器的控制逻辑部分。

（2）步骤 3、4、5：页面控制器接收到请求后，进行功能处理，首先需要收集和绑定请求参数到一个对象，这个对象在 Spring Web MVC 中称为命令对象，并进行验证，然后将命令对象委托给业务对象进行处理；处理完毕后返回一个 Model And View（模型数据和逻辑视图名）。

（3）步骤 6、7：前端控制器收回控制权，然后根据返回的逻辑视图名，选择相应的视图进行渲染，并把模型数据传入以便视图渲染。

（4）步骤 8：前端控制器再次收回控制权，将响应返回给用户。

ModelAndView 代码示例：

```
public ModelAndView getModelAndView() {
    ModelAndView view = new ModelAndView();
view.addObject("path","/");
    view.setViewName("/main");
    return view;
}
```

### 10. Spring 注解

注解是非常流行的技术，很多主流框架都支持注解，编写代码的时候尽量去用注解，注解使用方便，也可以使代码更加简洁。Spring 框架中也有很多注解，下面主要介绍项目中涉及的几类注解。

（1）@Controller：用于标注控制层组件。在 SpringMVC 中，控制器 Controller 负责处理由 DispatcherServlet 分发的请求，它把用户请求的数据经过业务处理层处理之后封装成一个 Model，然后再把该 Model 返回给对应的 View 进行展示。@Controller 对应表现层的 Bean，也就是 Action，用法如下。

```
@Controller
public class StorageController extends BaseController {}
```

（2）@Service：用于标注业务层组件，对应的是业务层 Bean，用法如下。

```
@ Service
public class SwiftStorageService  extends Storage {}
```

（3）@Repository：用于标注数据访问组件，即DAO组件，对应数据访问层Bean，用法如下。

```
@ Repository
public class ShareDao extends BaseDao<ShareBean> {}
```

（4）@Autowired：可以对类成员变量、方法及构造函数进行标注，完成自动装配的工作，用法如下。

```
@Autowired
private ShareService shareService;
```

（5）@RequestMapping：可以处理请求地址映射，用法如下。

```
@RequestMapping("/home")
public  ModelAndView  home(HttpServletRequest  request,  HttpServletResponse
response, String path) {
    ...
}
```

使用了@RequestMapping注解后，Spring收到URL请求http://localhost/home，将自动调用home()方法处理该请求。

（6）@ResponseBody：将返回类型直接输入到HTTP response body中，通常用于JSON格式数据的输出，用法如下。

```
@RequestMapping("/login")
@ResponseBody
public Object login(HttpServletRequest request, HttpServletResponse response,
String username, String password){
    return new MessageBean(true,Constants.SUCCESS_1);
}
```

在用Ajax异步提交数据的时候，后台对接的方法需要加上@ResponseBody注解。

### 3.3.2 实现步骤

#### 1. 实现控制层

利用Spring的框架实现页面之间的跳转，代码如下：

```
/**
    * 登录
    * @param request:请求
    */
   @RequestMapping("/getlogin")
   public ModelAndView login(HttpServletRequest request) {
       ModelAndView view=new ModelAndView();
```

```
        view.setViewName("/main");
        return view;
    }
```

### 2. 实现界面层

在前端页面之中设置一个 button 按钮，代码如下：

```
<body class="course-dashboard-page">
账号：<input id="username" >
密码：<input  id="password"><button class="button" onclick="getlogin()" >登入
</button>
    <script src="assets/javascripts/jquery.min.js"></script>
    <script type="text/javascript">
    function getlogin(){
        location.href="getlogin.action";
    }
    </script>
</body>
```

### 3. 功能验证测试

将项目部署到 Tomcat 中并发布，在浏览器之中输入 URL 地址，效果如图 3-10 所示。

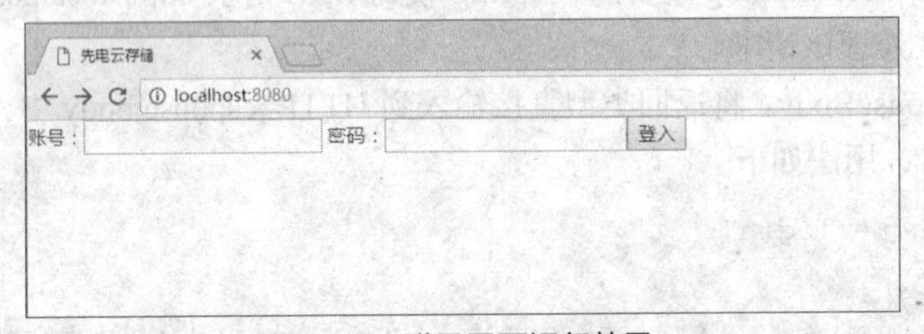


图 3-10　登录界面运行效果

输入账号和密码后，单击“登入”按钮后，切换到主界面，如图 3-11 所示。

图 3-11　登录后主界面

## 任务 3.4　了解 Hibernate 相关知识

### 3.4.1　相关知识

Hibernate 框架是一个开源的对象关系映射（ORM）框架，是对 JDBC 的轻量级的对象封装，使 Java 程序员可以使用面向对象的方式操控数据库。

Hibernate 是一种 ORM 框架。在使用 Hibernate 时，开发人员需要创建一个和表结构（R）相对应的 Java 类（O），然后通过 Hibernate 将这两者相映射（M），这样对表的操作就完全转换为对 Java 对象的操作。ORM 的原理如图 3-12 所示。

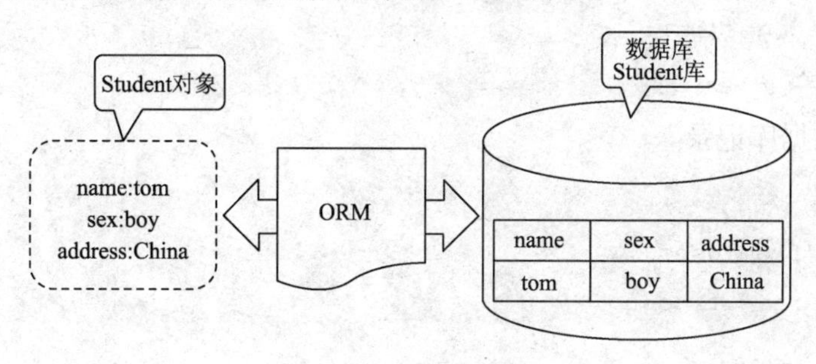


图 3-12　ORM 原理

Hibernate 框架介于数据库和应用之间，技术架构如图 3-13 所示。

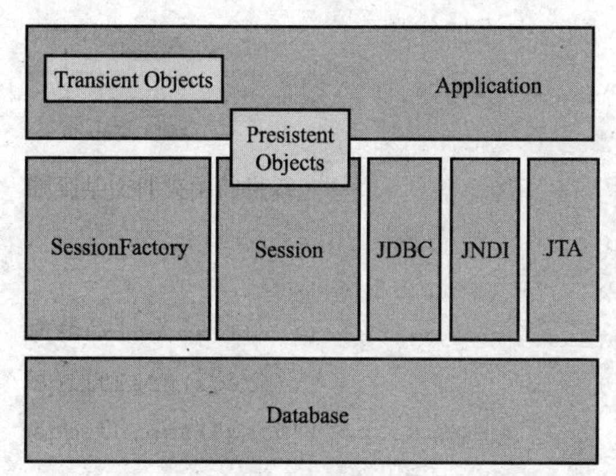


图 3-13　Hibernate 框架

Hibernate 框架内容比较多，此处不详细讨论，只介绍项目涉及的 HQL 相关技术。HQL(Hibernate Query Language)是面向对象的查询语言，它和 SQL 查询语言相似，在 Hibernate 提供的各种检索方式中，HQL 是使用最广的一种检索方式。

在本项目的开发中，通过引入 HibernateTemplate 完成对数据库增(save)、删(remove)、改(update)、查(find)的操作。使用 HibernateTemplate 非常简单，创建 HibernateTemplate 实例后，注入一个 SessionFactory 的引用，就可执行持久化操作。

Hibernate 的用法举例如下。

（1）增加数据库记录。

```
User user = new User();
user.setXXX();
```

```
dao.save(user);
```

（2）删除数据库记录。

```
//更新 id 获取用户
User user = dao.get(id);
dao.remove(user);
```

（3）更新操作记录。

```
//更新 id 获取用户
User user = dao.get(id);
dao.update(user);
```

（4）查询数据库记录。

```
//hql 查询语句
String hql = "from User";
List<User> users = find(hql);
```

### 3.4.2 实现步骤

#### 1. 实现对数据库的查询操作

（1）在 user 的 Dao 层之中添加代码。

```
public User getUserByname(String username)
    {
        String hql = "from User where username = ?";
        List<User> users = find(hql,username);
        if (users==null || users.size() == 0) {
            return null;
        }else{
            return users.get(0);
        }
    }
```

（2）在 user 的 service 之中添加代码。

```
@Autowired
private UserDao userDao;
/**
 * 根据用户名查找用户
 * @param mail
 * @return
 */
public User getUserByname(String username)
{
```

```
    return userDao.getUserByname(username);
}
```

（3）在控制层之中添加代码。

```
@Autowired
    private UserService userService;
    /**
     * 验证账号密码
     * @param request
     * @param username 用户名
     * @param password 密码
     * @return
     */
    @RequestMapping("/login")
    public Object login(HttpServletRequest request, String username,String
password) {
        User user=userService.getUserByname(username);
        if (user == null) {
            return new MessageBean(false,Constants.ERROR_1);
        }
        String inputstr = user.getPassword();
        if (!inputstr.equals(password)) {
            return new MessageBean(false,Constants.ERROR_2);
        }
        return new MessageBean(true,Constants.SUCCESS_1);
    }
```

（4）在前台增加交互代码。

```
function getlogin() {
            var username = $("#username").val();
            var password = $("#password").val();
            var data = {
                username : username,
                password : password
            };
            $.ajax({
                url : "login.action",
                type : "post",
                data : data,
                success : function(s) {
```

```
                if (s.success) {
                    alert(s.msg)
                    location.href = "getlogin.action";
                } else {
                    alert(s.msg);
                }
            }
        });
    }
```

### 2. 创建用户表

在 MySQL 中创建用户表“user”，并增加一条数据，用户名为“test”，密码为“123456”。

### 3. 功能验证测试

将项目部署到 Tomcat 中并发布，在浏览器之中输入 URL 地址，输入用户名“test”与密码“123456”，运行效果如图 3-14 所示。

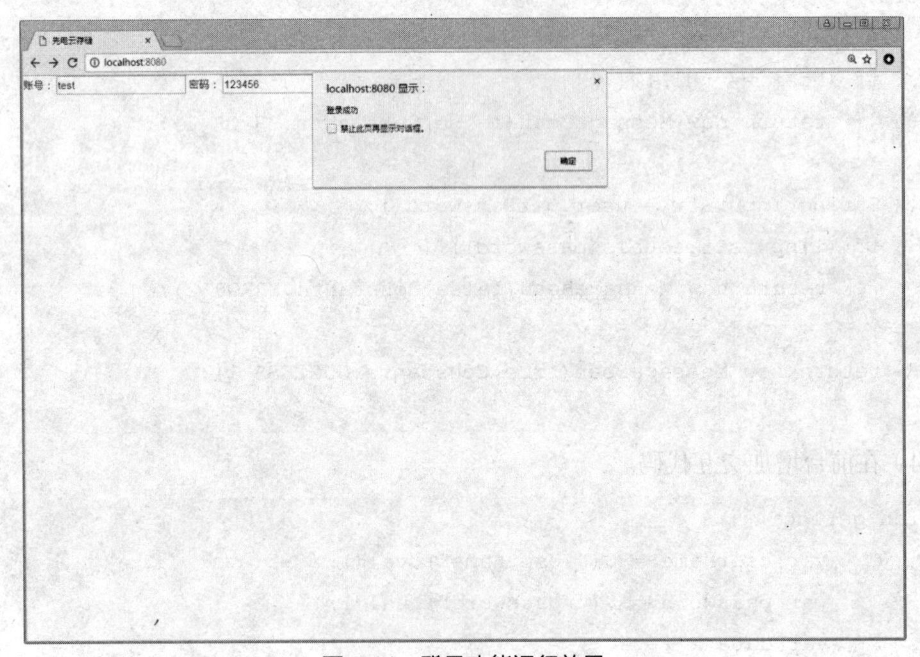


图 3-14　登录功能运行效果

# 项目 4 云存储 OpenStack Swift 服务构建

## 单元介绍

本单元读者需要学习 OpenStack Swift 的基本原理和内部使用的基本命令，掌握如何搭建 OpenStack Swift 服务，最后对其 APIs 进行测试。

## 学习任务

本单元主要完成以下学习目标：

- 掌握云存储服务 OpenStack Swift 的搭建步骤；
- 掌握 Swift 服务的使用；
- 掌握 Swift APIs 的使用；
- 掌握 OpenStack Swift SDK 的获取、编译和测试方法。

根据这些学习目标，本项目将分解为以下 3 个任务，见表 4-1。

表 4-1　任务分解表

| 任务序号 | 任务内容 | 验证方式 |
| --- | --- | --- |
| 任务 4.1 | 搭建 OpenStack Swift 服务 | 搭建服务并上传所需文件 |
| 任务 4.2 | Swift 服务 RESTful APIs 测试 | 使用 Swift 对文件进行操作 |
| 任务 4.3 | OpenStack Swift SDK 测试 | 对 OpenStack Swift SDK 获取编译和测试 |

## 任务 4.1　搭建 OpenStack Swift 服务

### 4.1.1　相关知识

#### 1. Swift 的基本概念

（1）Account

出于访问安全性考虑，使用 Swift 系统时，每个用户必须有一个账号（Account）。只有通过 Swift 验证的账号才能访问 Swift 系统中的数据。提供账号验证的节点称为 Account Server。Swift 由 Swauth 提供账号权限认证服务。

用户通过账号验证后将获得一个验证字符串（authentication token），后续的每次数据访问操作都需要传递这个字符串。

（2）Container

Swift 中的 Container 可以类比 Windows 操作系统中的文件夹或者 Unix 类操作系统中的目录，用于组织管理数据，所不同的是，Container 不能嵌套。数据都以 Object 的形式存放在 Container 中。

（3）Object

Object（对象）是 Swift 中的基本存储单元。一个对象包含两部分，数据和元数据（Metadata）。其中元数据包括对象所属 Container 名称、对象本身名称，以及用户添加的自定义数据属性（必须是 Key-Value 格式）。

对象名称在 URL 编码后大小要求小于 1024 字节。用户上传的对象最大是 5GB。用户可以通过 Swift 内建的大对象支持技术获取超过 5GB 的大对象。对象的元数据不能超过 90 个 Key-Value 对属性，并且这些属性的总大小不能超过 4KB。

Account、Container、Object 是 Swift 系统中的 3 个基本概念，三者的层次关系是：一个 Account 可以创建拥有任意多个 Container，一个 Container 中可以包含任意多个 Object。

在 Swift 系统中，集群被划分成多个区（Zone），每个区可以是一个磁盘，一个服务器，一台机柜甚至一个数据中心，每个区中有若干个节点（Node）。Swift 将 Object 存储在节点（Node）上，每个节点都是由多个硬盘组成的，并保证对象在多个节点上有备份（默认情况下，Swift 会给所有数据保存 3 个副本），以及这些备份之间的一致性。备份将均匀地分布在集群服务器上，并且系统保证各个备份分布在不同区的存储设备上，这样可以提高系统的稳定性和数据的安全性。它可以通过增加节点来线性地扩充存储空间。当一个节点出现故障，Swift 会从其他正常节点对出故障节点的数据进行备份。

### 2. Swift 服务的优势

（1）高数据持久性

Swift 提供多重备份机制，拥有极高的数据可靠性，数据存放在高分布式的 Swift 系统中，几乎不会丢失，Swift 在 5 个 Zone、5×10 个存储节点、数据复制 3 份时，数据持久性的 SLA 能够达到 10 个 9，即存储 1 万个文件到 Swift 中，经过 10 万年后，可能会丢失一个文件，这种文件丢失几乎可以忽略不计。

（2）极高的可拓展性

Swift 通过独立节点来形成存储系统。首先，Swift 在数据量的存储上就做到了无限拓展。另外，Swift 的性能也可以通过增加 Swift 集群来实现线性提升，所以 Swift 很难达到性能瓶颈。

（3）无单点故障

由于 Swift 的节点独立的特点，在实际工作时，不会发生传统存储系统的单点故障，传统系统即使通过 HA 来实现热备，但在主节点出现问题时，还是会影响整个存储系统的性能。而在 Swift 系统中，数据的元数据（Metadata）是通过 Ring 算法完全随机均匀分布的，且元数据也会保存多份，对于整个 Swift 集群而言，没有单点的角色存在。

（4）REST 架构

REST 是 Roy Fielding 博士在 2000 年他的博士论文中提出来的一种软件架构风格。

REST（Representational State Transfer）是一种轻量级的 Web Service 架构风格，其实现和操作比 SOAP 和 XML-RPC 更为简洁，可以完全通过 HTTP 协议实现，还可以利用 Cache 来提高响应速度，性能、效率和易用性上都优于 SOAP 协议。

REST 架构遵循了 CRUD 原则，CRUD 原则对于资源只需要 4 种行为：Create（创建）、Read（读取）、Update（更新）和 Delete（删除）就可以完成对其操作和处理。这 4 个操作是一种原子操作，即一种无法再分的操作，通过它们可以构造复杂的操作过程，正如数学上四则运算是数字的最基本的运算一样。

REST 架构对资源的操作包括获取、创建、修改和删除资源的操作正好对应 HTTP 协议提供的 GET、POST、PUT 和 DELETE 方法，因此 REST 把 HTTP 对一个 URL 资源的操作限制在 GET、POST、PUT 和 DELETE 这 4 个操作之内。这种针对网络应用的设计和开发方式，可以降低开发的复杂性，提高系统的可伸缩性。

因为其简洁方便性，越来越多的 Web 服务开始采用 REST 风格设计和实现。Swift 采用 REST 架构，不能像普通的文件系统那样对数据进行访问，必须通过它提供的 API 来访问操作数据，如图 4-1、图 4-2 和图 4-3 所示，分别展示了 Swift 的上传和下载操作。

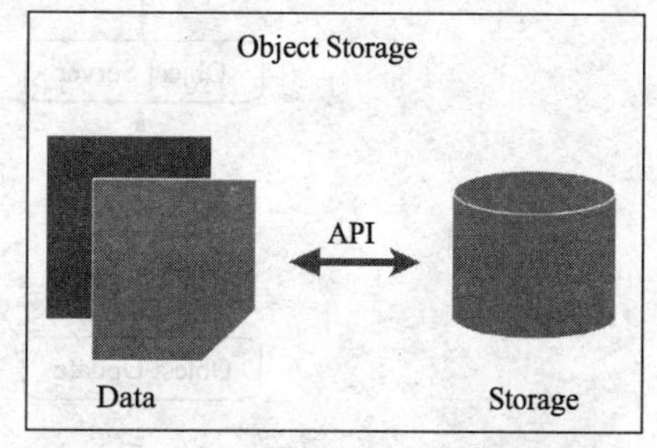

图 4-1　Swift API 访问存储的数据

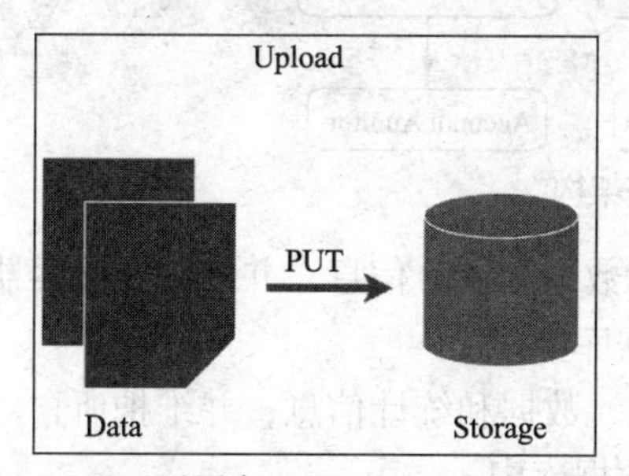

图 4-2　Swift 通过 HTTP PUT 方法上传数据

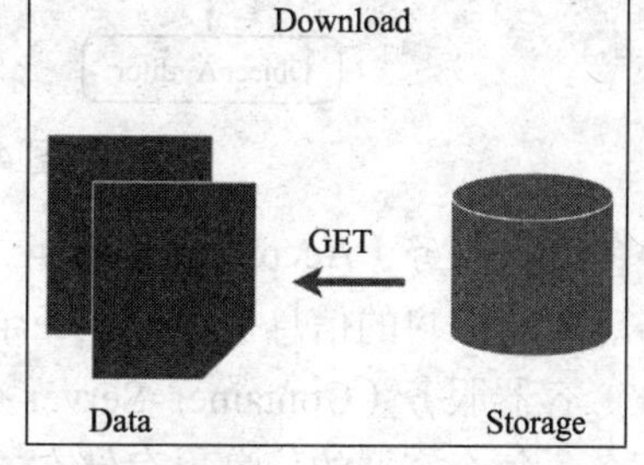

图 4-3　Swift 通过 HTTP GET 方法下载数据

Swift 对操作数据的 API 做了不同语言的封装绑定，以方便开发者进行开放。目前支持的语言有 PHP、Python、Java、C#/.NET 和 Ruby。

3. Swift 服务架构

Swift 集群主要包含认证节点、代理节点和存储节点。认证节点主要负责对用户的请求授权，只有通过认证节点授权的用户才能操作 Swift 服务，Swift 的技术架构如图 4-4 所示。

- 代理服务（Proxy Server）：对外提供对象服务 API，根据环的信息来查找服务地址并转发用户请求至相应的账户、容器或者对象服务；由于采用无状态的 REST 请求协议，可以进行横向扩展来均衡负载。
- 认证服务（Authentication Server）：验证访问用户的身份信息，并获得一个对象访问令牌（Token），在一定的时间内会一直有效；验证访问令牌的有效性并缓存下来直至过期时间。
- 缓存服务（Cache Server）：缓存的内容包括对象服务令牌，账户和容器的存在信息，但不会缓存对象本身的数据；缓存服务可采用 Memcached 集群，Swift 会使用一致性散列算法分配缓存地址。

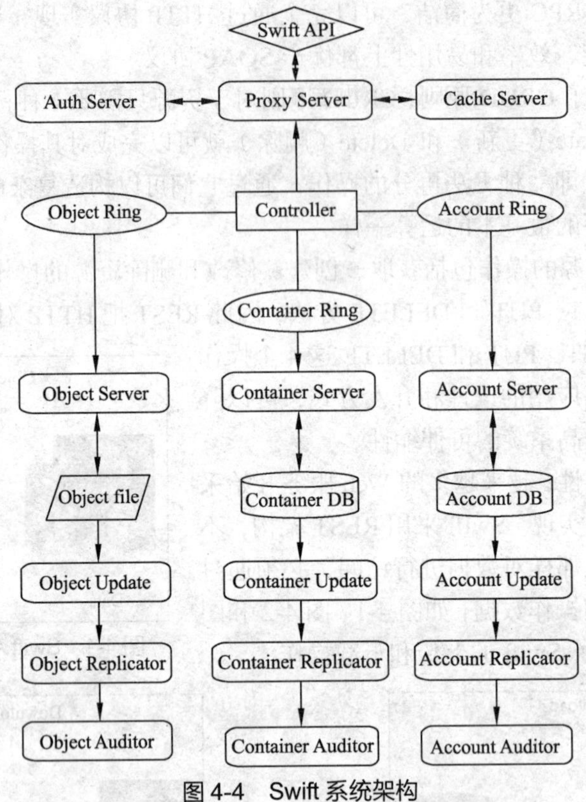


图 4-4　Swift 系统架构

- 账户服务（Account Server）：提供账户元数据和统计信息，并维护所含容器列表的服务，每个账户的信息被存储在一个 SQLite 数据库中。
- 容器服务（Container Server）：提供容器元数据和统计信息，并维护所含对象列表的服务，每个容器的信息也存储在一个 SQLite 数据库中。
- 对象服务（Object Server）：提供对象元数据和内容服务，每个对象的内容会以文件的形式存储在文件系统中，元数据会作为文件属性来存储，建议采用支持扩展属性的 XFS 文件系统。
- 复制服务（Replicator）：会检测本地分区副本和远程副本是否一致，具体是通过对比散列文件和高级水印来完成，发现不一致时会采用推式（Push）更新远程副本，例如，对象复制服务会使用远程文件复制工具 rsync 来同步；另外一个任务是确保被标记删除的对象从文件系统中移除。
- 更新服务（Updater）：当对象由于高负载的原因而无法立即更新时，任务将会被序列化到在本地文件系统中进行排队，以便服务恢复后进行异步更新。例如，成功创建对象后容器服务器没有及时更新对象列表，这个时候容器的更新操作就会进入排队中，更新服务会在系统恢复正常后扫描队列并进行相应的更新处理。
- 审计服务（Auditor）：检查对象、容器和账户的完整性，如果发现比特级的错误，文件将被隔离，并复制其他的副本以覆盖本地损坏的副本；其他类型的错误会被记录到日志中。

● 账户清理服务（Account Reaper）：移除被标记为删除的账户，删除其所包含的所有容器和对象。

● 环（Ring）：Ring 是 Swift 最重要的组件，用于记录存储对象与物理位置间的映射关系。在涉及查询 Account、Container、Object 信息时，就需要查询集群的 Ring 信息。Ring 使用 Zone、Device、Partition 和 Replica 来维护这些映射信息。Ring 中每个 Partition 在集群中都（默认）有 3 个 Replica。每个 Partition 的位置由 Ring 来维护，并存储在映射中。Ring 文件在系统初始化时创建，之后每次增减存储节点时，需要重新平衡一下 Ring 文件中的项目，以保证增减节点时，系统因此而发生迁移的文件数量最少。

● 区域（Zone）：Ring 中引入了 Zone 的概念，把集群的 Node 分配到每个 Zone 中。其中同一个 Partition 的 Replica 不能同时放在同一个 Node 上或同一个 Zone 内。防止造成所有的 Node 如果都在一个机架或一个机房中，一旦发生断电、网络故障等，造成用户无法访问的情况出现。

4. 基本命令

Swift 命令是用户管理 Swift 存储的接口，常用的命令列举如下。

（1）swift stat 命令。

功能：根据给定的参数显示账户、对象或容器的信息。

格式：

```
swift stat [container][object]
```

参数说明：

[container] 容器名称

[object] 对象名称

（2）swift list 命令。

功能：列出该账户的容器或容器的对象。

格式：

```
swift list [command-options] [container]
```

参数说明：

[command-options] 选项

[container] 容器名称

（3）swift upload 命令。

功能：根据参数将指定的文件或者目录上传到容器内。

格式：

```
swift upload [command-options] container file_or_directory [file_or_directory] [...]
```

参数说明：

[command-options] 选项

container 容器名称，或者是容器内的目录

file_or_directory 本地文件系统内的目录或者文件

[file_or_directory] 本地文件系统内的目录或者文件，可同时上传多个目录或文件

（4）swift post 命令。

功能：根据给定的参数升级 account、Container 或者 object 的元数据信息。

格式：

```
swift post [command-options] [container] [object]
```

参数说明：

[command-options] 选项

[container] 容器名称

[object] 对象名称

（5）swift download 命令。

功能：根据给定的参数下载容器中的对象。

格式：

```
swift download [command-options] [container] [object] [object] [...]
```

参数说明：

[command-options]选项

[container]容器名称

[object]对象名称(可同时下载多个对象)

（6）swift delete 命令。

功能：根据给定的参数删除容器中的对象。

格式：

```
swift delete [command-options] [container] [object] [object] [...]
```

参数说明：

[command-options] 选项

[container]容器名称

[object]对象名称(可同时删除多个对象)

### 4.1.2 实现步骤

本任务搭建单节点的 Swift 服务，需要安装 CentOS 6.5_x64 桌面操作系统，配置主机名，然后将提供的压缩包导入到该操作系统。

#### 1. 配置主机名

配置节点主机名为 Swift，配置完成通过如下命令验证：

```
$ vi /etc/sysconfig/network              //修改主机名和网络设置
NETWORKING=yes
HOSTNAME=Swift                           //修改主机名为 Swift （永久生效）
$ hostname  Swift                        // 临时修改系统主机名
```

```
$ hostname                              // 查询当前系统主机名
Swift
```

### 2. 配置环境

配置防火墙规则和 Selinux。

```
# 配置防火墙
# iptables -F  //清除所有 chains 链（INPUT/OUTPUT/FORWARD）中所有的 rule 规则
# iptables -Z  //清空所有 chains 链（INPUT/OUTPUT/FORWARD）中包及字节计数器
# iptables -X  //清除用户自定义的 chains 链（INPUT/OUTPUT/FORWARD）中的 rule 规则
# service iptables save //保存修改的 Iptables 规则
# 配置 selinux
修改配置文件  /etc/selinux/config
SELINUX=permissive      //表示系统会收到警告讯息但是不会受到限制，作为 selinux 的 debug
模式用处
# 保存修改内容后退出
```

### 3. 配置 YUM 源

将提供的安装光盘和安装文件复制到系统内部，制作安装源。本次测试采用实验室本地源。

移除系统 yum 源文件。

```
#  mv/etc/yum.repos.d/*/opt/
```

创建 repo 文件。

本次安装源为提供的教材光盘内的 iaas-repo 文件夹和 centos6.5 文件夹。

注：centos6.5 存放安装光盘的全部文件。

在/etc/yum.repos.d 创建 local.repo 源文件，搭建 ftp 服务器指向存放 yum 源路径：

```
[centos]
name=centos             //设置此 yum 的资源描述名称
baseurl=ftp://192.168.2.10/centos6.5/  //设置 yum 源的访问地址及路径
//（ 注：具体的 yum 源根据真实环境配置，本次为实验室测试环境 ）
gpgcheck=0              //禁用 gpg 检查 gpgkey
enabled=1               //启动此 yum 源
[OpenStack]
name=OpenStack
baseurl=ftp://192.168.2.10//iaas-repo/
//（ 注：具体的 yum 源根据真实环境配置，本次为实验室测试环境 ）
gpgcheck=0
enabled=1
# yum clean all         //清除缓存
```

### 4. 配置 IP

配置临时 IP，方便运行安装脚本，修改设备的 eth0 端口地址，修改配置文件 /etc/sysconfig/network-scripts/ifcfg-eth0 信息如下：

```
DEVICE=eth0                    //配置网卡的设备名称
IPADDR=172.24.0.10             //配置网络地址
NETMASK=255.255.255.0          //配置网络子网掩码
GATEWAY=172.24.0.1             //配置网络网关
BOOTPROTO=static               //配置静态网络地址
ONBOOT=yes                     //开机启动网络
USERCTL=no                     //不允许非 root 用户修改此设备
```

修改完成后，重启网络：

```
# service network restart
```

### 5. 重启设备

完成配置后，重启设备。

### 6. 部署脚本安装平台

将提供的安装脚本复制到主机中：

```
Xiandian_Pre.sh
Xiandian_Install_Controller_Node.sh
Xiandian_Install_Storage_Node.sh
```

### 7. 配置环境变量

修改 Xiandian_Pre.sh，内容如下：

```
MySQL_Admin_Passwd=000000                        //数据库用户密码
Admin_Passwd=000000                              //管理员密码
Demo_User_Passwd=000000                          //演示用户密码
Demo_DB_Passwd=000000                            //演示数据库密码
Contoller_Hostname=Swift                         //控制节点主机名
Controller_Mgmt_IPAddress=172.24.0.10            //控制节点管理网段密码
Gateway_Mgmt=172.24.0.1                          //管理网段网关
Controller_Stroage_IPAddress=172.24.1.10         //存储网络地址
Controller_External_IPAddress=172.24.2.10        //外部地址
Stroage_Hostname=Swift                           //存储节点主机名
Stroage_Mgmt_IPAddress=172.24.0.10               //存储节点管理地址
Stroage_Stroage_IPAddress=172.24.1.10            //存储节点存储地址
Stroage_External_IPAddress=172.24.2.10           //存储节点外部地址
Stroage_Swift_Disk=sda2                          //Swift 存储磁盘分区名称
```

修改完成之后保存配置并退出。

8. 配置控制节点

配置完成环境变量之后，控制节点执行“./Xiandian_Install_Controller_Node.sh”，在执行过程中敲击回车键完成密钥创建，同时输入节点密码完成密钥验证。执行完成之后再执行“./Xiandian_Install_Storage_Node.sh”，完成安装。

9. 验证 Swift 服务

执行以下命令，查看 Swift 服务状态是否正确：

```
# source /etc/keystone/admin-openrc.sh
# swift-init all restart
#swift stat
```

如果出现以下提示信息，表示 Swift 服务安装成功：

```
[root@Swift ~]# Swift stat
      Account: AUTH_fce45633eb824df690df8d8944faa15d
   Containers: 49
      Objects: 171
        Bytes: 761817242
Accept-Ranges: bytes
  X-Timestamp: 1446138807.29493
   X-Trans-Id: tx68e2f257f30142a585a66-00575dadfe
 Content-Type: text/plain; charset=utf-8
```

10. 开发环境构建

首先将 Swift-server.zip 解压到桌面。然后安装 Virtual Box 软件。在本书提供的软件包内有 Virtual Box 的安装包，用户也可以自行下载最新版本的 Virtual Box 安装包进行安装。

安装 Virtual Box 完成之后，打开本地物理 PC 网络连接，右键“VirtualBoxHost-Only Network”网卡，单击“属性”按钮，修改虚拟网卡配置，如图 4-5 所示。

将 IP 地址设置为“192.168.1.1”，子网掩码配置为“255.255.255.0”，然后单击“确定”按钮。

打开 Virtual Box 主界面，单击“新建”按钮，设置虚拟机参数如下。

- 名称：swift。
- 类型：Linux。
- 版本：Linux2.6/3.x/4.x(64-bit)。
- 内存大小：推荐 1GB 以上内存空间大小。
- 虚拟硬盘：使用已有的虚拟硬盘文件。

参数设置完成后，选择“虚拟硬盘”选项卡内的“使

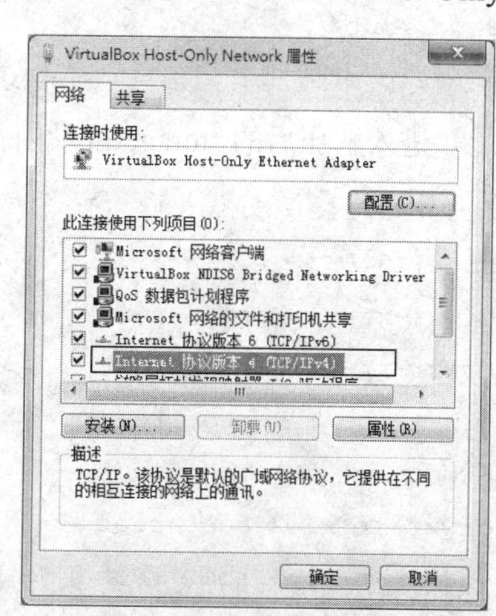


图 4-5　设置虚拟网卡属性

用已有的虚拟硬盘文件”选项，单击文件夹图标选择虚拟硬盘。

在新建虚拟机的过程当中，如果版本没有“Linux2.6/3.x/4.x（64-bit）”这一选项。可以先选择“Linux2.6/3.x/4.x(32-bit)”选项。

找到桌面 siwft-server/swift1 目录下的 vmdk 文件，选择 swift1.vmdk 文件，然后单击“创建”按钮。

创建完成后，右击创建好的 swift 虚拟机按钮，单击“设置”菜单项。

选择版本为“Linux2.6/3.x/4.x（64-bit）”选项。如果无法选择这一版本，需要重启物理 PC，然后进入 BIOS 界面开启 PC 的虚拟化支持。

单击“网络”标签，选择连接方式为“仅主机（Host-Only）适配器”，然后单击“启动”按钮来启动虚拟机。

虚拟机启动后，可以直接使用账号“admin”和密码“000000”登录虚拟机，也可以也可以通过 SecureCRT 连接虚拟机。

接下来创建用户和租户，并赋予租户操作 Swift 服务的权限：

```
# keystone user-create --name gw001 -pass 000000
# keystone tenant-create --name gw001
# keystone user-role-add --user gw001 --tenant gw001 --role SwiftOperator
```

用户建立完成后，可以得到连接 Swift 服务的参数为：

```
username: gw001
password: 000000
tenantname: gw001
```

至此 Swift 服务搭建完毕。

**11. 上传数据**

现在通过脚本在 Swift 服务中上传数据。将 storage.zip 上传到 Swift 服务器的/opt/目录下。将压缩包解压到/opt/目录：

```
# cd /opt/
# unzip storage.zip
```

进入解压后的 storage 目录内，可以查看到目录内的所有文件：

```
[root@Swift storage]# ll
total 344
drwxr-xr-x. 2 root root      4096 Jun 15 09:35 bigdata
drwxr-xr-x. 2 root root      4096 Jun 15 09:35 cloudskill
-rw-r--r--. 1 root root      9216 Jun 15 09:35 cloudskill.doc
-rw-r--r--. 1 root root     15038 Jun 15 09:35 cloudskill.png
drwxr-xr-x. 2 root root      4096 Jun 15 09:35 iaas
-rw-r--r--. 1 root root      7680 Jun 15 09:35 inbigdata.ppt
-rw-r--r--. 1 root root        23 Jun 15 09:35 incloudskill.txt
```

```
-rw-r--r--. 1 root root      7168 Jun 15 09:35 iniaas.xls
-rw-r--r--. 1 root root        17 Jun 15 09:35 inpaas.txt
-rw-r--r--. 1 root root       168 Jun 15 09:35 insaas.txt
-rw-r--r--. 1 root root    243104 Jun 15 09:35 inxdcloud.mp4
drwxr-xr-x. 2 root root      4096 Jun 15 09:35 paas
-rw-r--r--. 1 root root      3343 Jun 15 09:35 products.jpg
-rw-r--r--. 1 root root      8065 Jun 15 09:35 products.png
drwxr-xr-x. 2 root root      4096 Jun 15 09:35 saas
-rwxr-xr-x. 1 root root      2331 Jun 15 09:35 swiftupload.sh
drwxr-xr-x. 2 root root      4096 Jun 15 09:35 tmp
drwxr-xr-x. 2 root root      4096 Jun 15 09:35 xdcloud
drwxr-xr-x. 2 root root      4096 Jun 15 09:35 四大名著
```

通过 cat 命令查看 swiftupload.sh 文件：

```
[root@swift storage]# cat swiftupload.sh
#!/bin/bash
```

定义上传对象 uploadFile 脚本：

```
uploadFile( )
{
#删除容器，并创建新的容器
echo "################## Delete Container #################"
swift delete gw001
echo "################## Create Container #################"
swift post gw001
#上传目录对象，完成后将文件复制到目录内，然后上传文件对象
#文件对象上传完成后删除目录内的文件
swift   upload  gw001  bigdata/
cp inbigdata.ppt bigdata/
swift   upload  gw001  bigdata/inbigdata.ppt
rm -rf bigdata/inbigdata.ppt

swift   upload  gw001  iaas/
cp iniaas.xls iaas/
swift   upload  gw001  iaas/iniaas.xls
rm -rf iaas/iniaas.xls

swift   upload  gw001  paas/
cp inpaas.txt paas/
swift   upload  gw001  paas/inpaas.txt
```

```
rm -rf paas/inpaas.txt

Swift   upload  gw001  saas/
cp insaas.txt  saas/
swift   upload  gw001  saas/insaas.txt
rm -rf saas/insaas.txt

swift upload gw001 cloudskill/
cp incloudskill.txt cloudskill/
swift upload gw001 cloudskill/incloudskill.txt
rm -rf cloudskill/incloudskill.txt

swift upload gw001 xdcloud/
cp inxdcloud.mp4 xdcloud/
swift upload gw001 xdcloud/inxdcloud.mp4
rm -rf xdcloud/inxdcloud.mp4

#上传四大名著目录对象
swift upload gw001 四大名著/

#在目录内新建 doc、pdf、txt 三个目录
mkdir 四大名著/doc
mkdir 四大名著/pdf
mkdir 四大名著/txt

#上传 doc、pdf、txt 三个目录对象
swift upload gw001 四大名著/doc/
swift upload gw001 四大名著/pdf/
swift upload gw001 四大名著/txt/

#将 tmp 目录内的文件分类复制到 doc、pdf、txt 目录内
cp tmp/*.docx 四大名著/doc/
cp tmp/*.pdf 四大名著/pdf/
cp tmp/*.txt 四大名著/txt/

#上传 doc、pdf、txt 目录内的文件对象
swift upload gw001 四大名著/doc/
swift upload gw001 四大名著/pdf/
swift upload gw001 四大名著/txt/
```

```
#删除四大名著目录内的所有内容
rm -rf 四大名著/*

#上传文件对象到容器的根目录
swift   upload gw001 cloudskill.png
swift   upload gw001 cloudskill.doc
swift   upload gw001 products.jpg
swift   upload gw001 products.png

echo "###################################################"
echo "###################   FileList   #################"
echo "###################################################"

#展示容器内的文件列表
swift list gw001
}
```

通过脚本注释可以看出，执行脚本首先会删除并创建一个 gw001 的容器，然后依次将 Storage 目录中的内容上传到 gw001 的容器内。

上传之前，需要切换到 gw001 用户登录。

```
# export OS_USERNAME=admin
# export OS_PASSWORD=000000
# export OS_TENANT_NAME=admin
# export OS_AUTH_URL=http://Swift:35357/v2.0
```

也可以将上述命令行写成 gw001-openrc.sh 文件。通过 source 命令快速切换到 gw001 用户。

```
# cat /opt/gw001-openrc.sh
 export OS_USERNAME=gw001
 export OS_PASSWORD=000000
 export OS_TENANT_NAME=gw001
 export OS_AUTH_URL=http://Swift:35357/v2.0
#source /opt/gw001-openrc.sh
```

用户切换完成后，执行 swiftupload.sh 文件，创建 gw001 容器内的数据。

```
# cd /opt/storage
# chmod +x swiftupload.sh
# ./swiftupload.sh
```

执行完成后，可以通过命令查看容器里的文件。

```
# swift list gw001
bigdata/
bigdata/inbigdata.ppt
cloudskill.doc
cloudskill.png
cloudskill/
cloudskill/incloudskill.txt
iaas/
iaas/iniaas.xls
paas/
paas/inpaas.txt
products.jpg
products.png
saas/
saas/insaas.txt
xdcloud/
xdcloud/inxdcloud.mp4
四大名著/
四大名著/doc/
四大名著/doc/西游记.docx
四大名著/pdf/
四大名著/pdf/红楼梦.pdf
四大名著/txt/
四大名著/txt/三国演义.txt
四大名著/txt/水浒传.txt
```

显示以上内容表明数据已经创建成功，这样 Swift 开发环境就全部搭建完成了。

## 任务 4.2 Swift 服务 RESTful APIs 测试

### 4.2.1 相关知识

API（Application Programming Interface，应用程序编程接口）是一些预先定义的函数，目的是提供应用程序与开发人员访问某种服务和资源的能力，而又无需访问源码，或理解内部工作机制的细节。

在介绍 API 程序之前，首先了解一下 Curl 工具的使用。Curl 是一个命令行工具，能够通过命令行发送和接受 HTTP 请求和响应，这使得它能够直接调用 REST API 进行工作。Curl 的主要命令见表 4-2。

表 4-2　Curl 主要命令

| -H <line> | 自定义头信息传递给服务器 |
| --- | --- |
| -i | 输出时包括 protocol 头信息，显示响应头 |
| -k | 允许不使用证书到 SSL 站点 |
| -v | 显示详细信息 |
| -X <command> | 指定命令 |
| -d <data> | HTTP POST 方式传送数据 |

Swift 通过 Proxy Server 向外提供基于 HTTP 的 REST 服务接口，对账户、容器和对象进行 CRUD 等操作。在访问 Swift 服务之前，需要先通过认证服务获取访问令牌，然后在发送的请求中加入头部信息 X-Auth-Token。下面是请求返回账户中的容器列表的示例。

首先需要获取用户的请求 token 值，如下所示：

```
# keystone_token=` Curl  -d '{"auth":{"tenantName":"admin","passwordCredentials":
{"username":"admin","password":"000000"}}}'  -H  "Content-type:application/json"
http://172.24.0.10:35357/v2.0/tokens | sed -e 's/"/ /g' -e  's/,/ /g' `
```

根据反馈的结果取出 token 值，赋予变量为 token

```
# token='echo $keystone_token |awk {'print $16'}'
```

取出 storage_url 地址，赋予变量为 storage_url

```
# publiCurl='echo $keystone_token |awk {'print $294'}'
```

根据以上用例，可以对照 Swift 的 RESTful API 表总结一下常用的命令，见表 4-3。

表 4-3　Swift RESTful API 说明

| 资源类型 | URL | GET | PUT | POST | DELETE | HEAD |
| --- | --- | --- | --- | --- | --- | --- |
| 账户 | /account/ | 获取容器列表 | | | | 获取账户元数据 |
| 容器 | /account/Container | 获取对象列表 | 创建容器 | 更新容器元数据 | 删除容器 | 获取容器元数据 |
| 对象 | /account/Container/object | 获取对象内容和元数据 | 创建、更新或复制对象 | 更新对象元数据 | 删除对象 | 获取对象元数据 |

## 4.2.2　实现步骤

### 1. 显示账号内容

格式：`GET  /v1/{account}`

用法：

```
# Curl -i $publiCurl?format=json -X GET -H "X-Auth-Token: $token"
```

```
HTTP/1.1 200 OK
Content-Length: 221
Accept-Ranges: bytes
X-Timestamp: 1457506647.35213
X-Account-Bytes-Used: 1051537
X-Account-Container-Count: 4
Content-Type: application/json; charset=utf-8
X-Account-Object-Count: 22
X-Trans-Id: tx208095d71df24926b4eeb-00573042b6
Date: Mon, 09 May 2016 07:56:38 GMT
[{"count": 0, "bytes": 0, "name": "BS_Dept_Private"}, {"count": 0, "bytes": 0,
"name": "IT_Dept_Private"}, {"count": 0, "bytes": 0, "name": "RD_Dept_Public"},
{"count": 22, "bytes": 1051537, "name": "Volume_test_backup"}]
```

2. 创建、更新、删除账号数据

格式：`POST /v1/{account}`

用法：

（1）通过创建的账号列举容器的内容。

```
# Curl -i $publiCurl -X POST -H "X-Auth-Token: $token" -H "X-Account-Meta-Book:
MobyDick" -H "X-Account-Meta-Subject: Literature"
HTTP/1.1 204 No Content
Content-Length: 0
Content-Type: text/html; charset=UTF-8
X-Trans-Id: txdfb91c5374ad4ce8a3800-00573043e8
Date: Mon, 09 May 2016 08:01:44 GMT
```

可以通过此账号显示相关信息：

```
# Curl -i $publiCurl?format=json -X GET -H "X-Auth-Token: $token"
HTTP/1.1 200 OK
Content-Length: 221
X-Account-Object-Count: 22
X-Account-Meta-Book: MobyDick
X-Timestamp: 1457506647.35213
X-Account-Meta-Subject: Literature
X-Account-Bytes-Used: 1051537
X-Account-Container-Count: 4
Content-Type: application/json; charset=utf-8
Accept-Ranges: bytes
X-Trans-Id: tx2cb18c35b09f44808fa7e-0057304545
```

```
Date: Mon, 09 May 2016 08:07:33 GMT
[{"count": 0, "bytes": 0, "name": "BS_Dept_Private"}, {"count": 0, "bytes": 0,
"name": "IT_Dept_Private"}, {"count": 0, "bytes": 0, "name": "RD_Dept_Public"},
{"count": 22, "bytes": 1051537, "name": "Volume_test_backup"}]
```

可以显示出 **X-Account-Meta-Subject** 的定义说明，接下来更新和修改这个内容。

（2）更新账号内容。

```
# Curl -i $publiCurl -X POST -H "X-Auth-Token: $token" -H "X-Account-Meta-Subject:
Xiandian_Swift"
HTTP/1.1 204 No Content
Content-Length: 0
Content-Type: text/html; charset=UTF-8
X-Trans-Id: txb9a3c10090f1409c94695-00573045dd
Date: Mon, 09 May 2016 08:10:05 GMT
```

此时再次查看详情：

```
# Curl -i $publiCurl?format=json -X GET -H "X-Auth-Token: $token"
HTTP/1.1 200 OK
Content-Length: 221
X-Account-Object-Count: 22
X-Account-Meta-Book: MobyDick
X-Timestamp: 1457506647.35213
X-Account-Meta-Subject: Xiandian_Swift
X-Account-Bytes-Used: 1051537
X-Account-Container-Count: 4
Content-Type: application/json; charset=utf-8
Accept-Ranges: bytes
X-Trans-Id: tx163e64e01cd44fa08babb-0057304606
Date: Mon, 09 May 2016 08:10:47 GMT
[{"count": 0, "bytes": 0, "name": "BS_Dept_Private"}, {"count": 0, "bytes": 0,
"name": "IT_Dept_Private"}, {"count": 0, "bytes": 0, "name": "RD_Dept_Public"},
{"count": 22, "bytes": 1051537, "name": "Volume_test_backup"}]
```

可以显示出 X-Account-Meta-Subject 已经被修改，接下来删除这个内容。

（3）删除账号内容。

```
# Curl -i $publiCurl -X POST -H "X-Auth-Token: $token" -H "X-Remove-Account-
Meta-Subject: x"
HTTP/1.1 204 No Content
Content-Length: 0
Content-Type: text/html; charset=UTF-8
```

```
X-Trans-Id: tx7f9859f168fb49eca65c5-0057304651
Date: Mon, 09 May 2016 08:12:01 GMT
```

删除完毕之后，检查删除内容：

```
# Curl -i $publiCurl?format=json -X GET -H "X-Auth-Token: $token"
HTTP/1.1 200 OK
Content-Length: 221
X-Account-Object-Count: 22
X-Account-Meta-Book: MobyDick
X-Timestamp: 1457506647.35213
X-Account-Bytes-Used: 1051537
X-Account-Container-Count: 4
Content-Type: application/json; charset=utf-8
Accept-Ranges: bytes
X-Trans-Id: txe019793cd2f7499ea04f2-0057304668
Date: Mon, 09 May 2016 08:12:24 GMT
[{"count": 0, "bytes": 0, "name": "BS_Dept_Private"}, {"count": 0, "bytes": 0,
"name": "IT_Dept_Private"}, {"count": 0, "bytes": 0, "name": "RD_Dept_Public"},
{"count": 22, "bytes": 1051537, "name": "Volume_test_backup"}]
```

此时发现已经删除了创建的账号。

### 3. 查看账号内容

格式：`POST    /v1/{account}`

用法：

```
# Curl -i $publiCurl -X HEAD -H "X-Auth-Token: $token"
HTTP/1.1 204 No Content
Content-Length: 0
X-Account-Object-Count: 22
X-Account-Meta-Book: MobyDick
X-Timestamp: 1457506647.35213
X-Account-Bytes-Used: 1051537
X-Account-Container-Count: 4
Content-Type: text/plain; charset=utf-8
Accept-Ranges: bytes
X-Trans-Id: tx89cc5492f3fd4236bef46-00573047af
Date: Mon, 09 May 2016 08:17:51 GMT
```

### 4. 查看容器内容、列举对象

格式：`Get    /v1/{account}/{Container}`

用法：

```
# Curl -i $publiCurl?format=json -X GET -H "X-Auth-Token: $token"
HTTP/1.1 200 OK
Content-Length: 221
X-Account-Object-Count: 22
X-Account-Meta-Book: MobyDick
X-Timestamp: 1457506647.35213
X-Account-Bytes-Used: 1051537
X-Account-Container-Count: 4
Content-Type: application/json; charset=utf-8
Accept-Ranges: bytes
X-Trans-Id: txe019793cd2f7499ea04f2-0057304668
Date: Mon, 09 May 2016 08:12:24 GMT
[{"count": 0, "bytes": 0, "name": "BS_Dept_Private"}, {"count": 0, "bytes": 0,
"name": "IT_Dept_Private"}, {"count": 0, "bytes": 0, "name": "RD_Dept_Public"},
{"count": 22, "bytes": 1051537, "name": "Volume_test_backup"}]
```

解析返回的 JSON 格式代码，可以查到容器内容，如下所示。

```
[
    {
        "count": 0,
        "bytes": 0,
        "name": "BS_Dept_Private"
    },
    {
        "count": 0,
        "bytes": 0,
        "name": "IT_Dept_Private"
    },
    {
        "count": 0,
        "bytes": 0,
        "name": "RD_Dept_Public"
    },
    {
        "count": 22,
        "bytes": 1051537,
        "name": "Volume_test_backup"
    }
]
```

5. 创建容器

格式：PUT    /v1/{account}/{Container}

用法：

```
# Curl -i $publiCurl/xiandian -X PUT -H "X-Auth-Token: $token" -H "X-Container-
Meta-Book: MobyDick"
HTTP/1.1 201 Created
Content-Length: 0
Content-Type: text/html; charset=UTF-8
X-Trans-Id: txb26b66b40a6c4d11964e4-0057304bf1
Date: Mon, 09 May 2016 08:36:01 GMT
```

查询当前容器列表：

```
# Curl -H "X-Auth-Token:$token" $publiCurl
BS_Dept_Private
IT_Dept_Private
RD_Dept_Public
Volume_test_backup
xiandian
```

6. 删除容器

格式：DELETE    /v1/{account}/{Container}

用法：

```
# Curl -i $publiCurl/xiandian -X DELETE -H "X-Auth-Token: $token"
HTTP/1.1 204 No Content
Content-Length: 0
Content-Type: text/html; charset=UTF-8
X-Trans-Id: tx820234e2d3854eb2a0b8e-0057304ce3
Date: Mon, 09 May 2016 08:40:03 GMT
```

查询当前容器列表：

```
# Curl -H "X-Auth-Token:$token" $publiCurl
BS_Dept_Private
IT_Dept_Private
RD_Dept_Public
Volume_test_backup
```

7. 创建、更新、删除容器数据

格式：POST    /v1/{account}/{Container}

用法：

（1）创建容器数据。

```
# Curl -i $publiCurl/xiandian -X POST -H "X-Auth-Token: $token" -H "X-Container-
Meta-Author:     MarkTwain"     -H     "X-Container-Meta-Web-Directory-Type:
text/directory" -H "X-Container-Meta-Century: Nineteenth"
```

（2）更新容器数据。

```
# Curl -i $publiCurl/xiandian -X POST -H "X-Auth-Token: $token" -H "X-Container-
Meta-Author: xiandian_Swift"
HTTP/1.1 204 No Content
Content-Length: 0
Content-Type: text/html; charset=UTF-8
X-Trans-Id: tx27805450d85a44e5a3b3a-005730573a
Date: Mon, 09 May 2016 09:24:10 GMT
```

（3）删除容器数据。

```
# Curl -i $publiCurl/xiandian -X POST -H "X-Auth-Token: $token" -H "X-Remove-
Container-Meta-Century: x"
HTTP/1.1 204 No Content
Content-Length: 0
Content-Type: text/html; charset=UTF-8
X-Trans-Id: tx3b1c927363274ced8b92f-0057305768
Date: Mon, 09 May 2016 09:24:56 GMT
```

### 8. 查看容器 Metadata 的元数据信息

格式：`HEAD    /v1/{account}/{Container}`

用法：查看 BS_Dept_Private 容器的元数据信息：

```
# Curl -i $publiCurl/BS_Dept_Private -X HEAD -H "X-Auth-Token: $token"
HTTP/1.1 204 No Content
Content-Length: 0
X-Container-Object-Count: 0
Accept-Ranges: bytes
X-Container-Meta-Century: Nineteenth
X-Timestamp: 1461118777.22888
X-Container-Meta-Author: MarkTwain
X-Container-Bytes-Used: 0
X-Container-Meta-Web-Directory-Type: text/directory
Content-Type: text/plain; charset=utf-8
X-Trans-Id: txef0bc7e765e54b1b9d00b-0057312fad
Date: Tue, 10 May 2016 00:47:41 GMT
```

接下来是关于 object 的 API 操作。环境存在两个 Container，一个容器名称为 xiandian，另一个为 android，两个容器内都存在一个 object。

9. 获取对象列表数据

格式：GET    /v1/{account}/{Container}/{object}

用法：

```
# Curl -i $publiCurl/xiandian/Swift -X GET -H "X-Auth-Token: $token"
HTTP/1.1 200 OK
Content-Length: 0
Accept-Ranges: bytes
Last-Modified: Tue, 10 May 2016 00:55:19 GMT
Etag: d41d8cd98f00b204e9800998ecf8427e
X-Timestamp: 1462841718.06502
Content-Type: application/octet-stream
X-Trans-Id: txbaf418f00a4d4bdc8e3d5-005731319a
Date: Tue, 10 May 2016 00:55:54 GMT
```

10. 创建或替代对象

格式：PUT    /v1/{account}/{Container}/{object}

用法：

```
# Curl -i $publiCurl/xiandian/helloworld -X PUT -H "Content-Length: 0" -H
"X-Auth-Token: $token"
HTTP/1.1 201 Created
Last-Modified: Tue, 10 May 2016 01:07:55 GMT
Content-Length: 0
Etag: d41d8cd98f00b204e9800998ecf8427e
Content-Type: text/html; charset=UTF-8
X-Trans-Id: txb84867db14ae4a0fa3d7f-005731346a
Date: Tue, 10 May 2016 01:07:54 GMT
```

11. 复制对象

格式：COPY    /v1/{account}/{Container}/{object}

用法：

```
# Curl -i $publiCurl/xiandian/helloworld -X COPY -H "X-Auth-Token: $token" -H
"Destination: android/hello"
HTTP/1.1 201 Created
Content-Length: 0
X-Copied-From-Last-Modified: Tue, 10 May 2016 01:07:55 GMT
X-Copied-From: xiandian/helloworld
Last-Modified: Tue, 10 May 2016 01:22:31 GMT
```

```
Etag: d41d8cd98f00b204e9800998ecf8427e
Content-Type: text/html; charset=UTF-8
X-Trans-Id: tx7d3f16b415d34996a9c02-00573137d5
Date: Tue, 10 May 2016 01:22:30 GMT
```

可以查看容器 android 下的对象列表：

```
# swift list android
hello
Swift
```

### 12. 删除对象

格式：DELETE /v1/{account}/{Container}/{object}

用法：

```
# Curl -i $publiCurl/xiandian/helloworld -X DELETE -H "X-Auth-Token: $token"
HTTP/1.1 204 No Content
Content-Length: 0
Content-Type: text/html; charset=UTF-8
X-Trans-Id: tx33ce1bcde631404bad728-005731389a
Date: Tue, 10 May 2016 01:25:46 GMT
# Swift list xiandian
Swift
```

### 13. 查看对象元数据

格式：HEAD /v1/{account}/{Container}/{object}

用法：

```
# Curl -i $publiCurl/xiandian/Swift -X HEAD -H "X-Auth-Token: $token"
HTTP/1.1 200 OK
Content-Length: 0
Accept-Ranges: bytes
Last-Modified: Tue, 10 May 2016 01:07:14 GMT
Etag: d41d8cd98f00b204e9800998ecf8427e
X-Timestamp: 1462842433.87203
Content-Type: application/octet-stream
X-Trans-Id: tx88d3ff38392c4e609b7d6-00573138df
Date: Tue, 10 May 2016 01:26:55 GMT
```

### 14. 创建或更新对象元数据

格式：POST /v1/{account}/{Container}/{object}

用法：

（1）创建对象元数据。

```
# Curl -i $publiCurl/xiandian/Swift -X POST -H "X-Auth-Token: $token" -H HTTP/1.1
202 Accepted
Content-Length: 76
Content-Type: text/html; charset=UTF-8
X-Trans-Id: tx8a37942419794e718f02f-0057313fac
Date: Tue, 10 May 2016 01:55:57 GMT
<html><h1>Accepted</h1><p>The request is accepted for processing.</p></html>
```

（2）更新元数据。

```
# Curl -i $publiCurl/xiandian/Swift -X POST -H "X-Auth-Token: $token" -H
"X-Object-Meta-Book: GoodbyeOldFriend"
HTTP/1.1 202 Accepted
Content-Length: 76
Content-Type: text/html; charset=UTF-8
X-Trans-Id: tx412c51736b904b589009f-005731409b
Date: Tue, 10 May 2016 01:59:55 GMT
```

## 任务 4.3 OpenStack Swift SDK 测试

### 4.3.1 相关知识

前面是对云存储服务、云存储 RESTful API 的学习，下面要开始网盘软件的开发了。首先选择支持 HTTP RESTful API 的开源 SDK，然后采用编译源码的方式进行代码测试，以了解 API 调用的基本机制。

目前 OpenStack SDK 有多种语言，其中支持 Java 的 SDK 有以下几种。

（1）Apache jclouds：是开源 Java 的云平台开发工具包，目前支持 AWS、Rackspace、OpenStack、CloudStack 等多个云平台的接入和操作，有丰富的文档。项目地址：http://jclouds.apache.org/。

（2）OpenStack4j：OpenStack 的 Java 客户端接入的 SDK，支持 OpenStack 认证、计算、镜像、网络、块存储、监控、数据处理等多个模块的接口操作。项目地址：http://www.OpenStack4j.com/。

（3）OpenStack-Java-SDK：也是 OpenStack 的 SDK，近一年更新较少，文档资源相对缺少。项目地址：https://github.com/woorea/OpenStack-Java-sdk。

（4）User Registration Service：基于 OpenStack-Java-sdk 开发的服务端的用户注册服务。

（5）JOSS（Java OpenStack Storage）：基于 OpenStack-Java-SDK 开发的 Swift 云存储的 APIs SDK。

本书选择 JOSS 作为 Java Web 云存储客户端的 SDK，编译成 jar 库文件，并参考 OpenStack Integration 的实现来构建 Swift 云存储客户端。

进入 JOSS 官网的帮助文档 http://joss.javaswift.org/，该文档介绍如何编译、引入 Java Swift SDK 库。本案例选择非 Maven 用户的单个 Jar 文件进行开发，这样可以避免引入相关依赖 Jar 文件的操作。到网站 https://github.com/JavaSwift/joss 下载源代码包，然后进行编译，如图 4-6 所示。

图 4-6　下载源代码包

## 4.3.2　实现步骤

在 Eclipse 工程之中新建一个 Test.java，用这个 Java 类来负责测试验证与 Swift 服务器对接是否成功，代码如下：

```
package test;
import org.JavaSwift.joss.client.factory.AccountConfig;
import org.JavaSwift.joss.client.factory.AccountFactory;
import org.JavaSwift.joss.model.Account;
import com.xiandian.cloud.storage.sh.security.QEncodeUtil;
import com.xiandian.cloud.storage.sh.util.SwiftUtilTools;
public class Test {
    public static void main(String[] args) {
        AccountConfig config = new AccountConfig();
        config.setUsername("admin");
        config.setPassword("XiandianSwift");
        config.setAuthUrl("http://58.214.31.6:5000/v2.0/tokens");
        config.setTenantName("admin");
        Account account = new AccountFactory(config).createAccount();
        System.out.println(account);
    }
}
```

# 项目 5　开发登录注册模块

## 单元介绍

本单元读者需掌握网盘登录与注册功能的实现，包括前后台代码的流程，并对其功能进行验证测试。

## 学习任务

本单元主要完成以下学习目标：

- 掌握网盘的登录功能和注册功能。

根据这些学习目标，本项目将分解为以下 2 个任务，见表 5-1。

表 5-1　任务分解表

| 任务序号 | 任务内容 | 验证方式 |
|---|---|---|
| 任务 5.1 | 开发登录功能 | 使用用户名和密码登录 |
| 任务 5.2 | 开发注册功能 | 注册新的用户名和密码 |

## 任务 5.1　开发登录功能

### 5.1.1　相关知识

本功能实现的技术原理如下：

（1）视图层：设计如图 5-1 所示的登录界面。

图 5-1　登录界面

（2）控制层：接收视图层的消息，向服务层发送登录信息，并将服务层返回的结果发送到视图层。

（3）服务层：调用 Keystone 服务对用户名和密码进行认证，并将认证结果返回给控制层。

本功能的具体实现流程如图 5-2 所示。

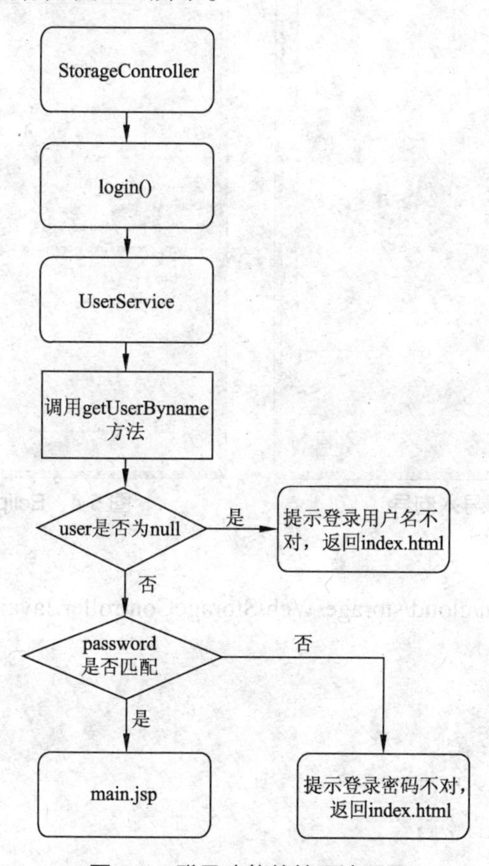


图 5-2　登录功能的处理流程图

### 5.1.2　实现步骤

#### 1. 导入项目

运行 Eclipse，选择“File”→“Import”命令，如图 5-3 所示。

选择“Existing Project into Workspace”选项，进入导入工程界面，如图 5-4 所示，单击“Browser”按钮导入工程 Project51。

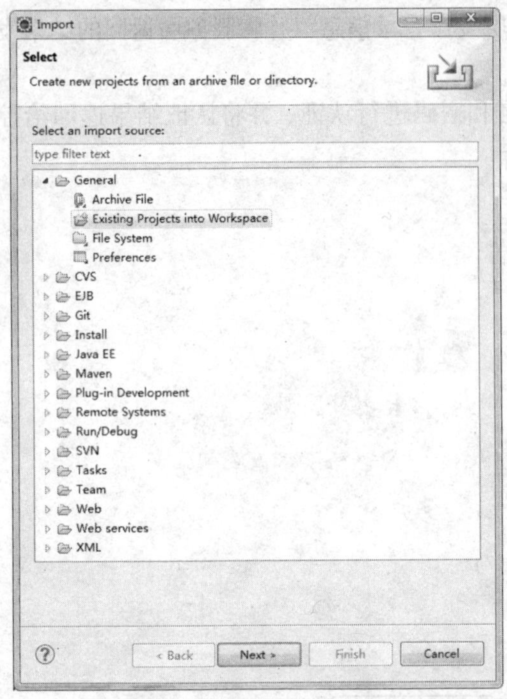


图 5-3　Eclipse 导入向导

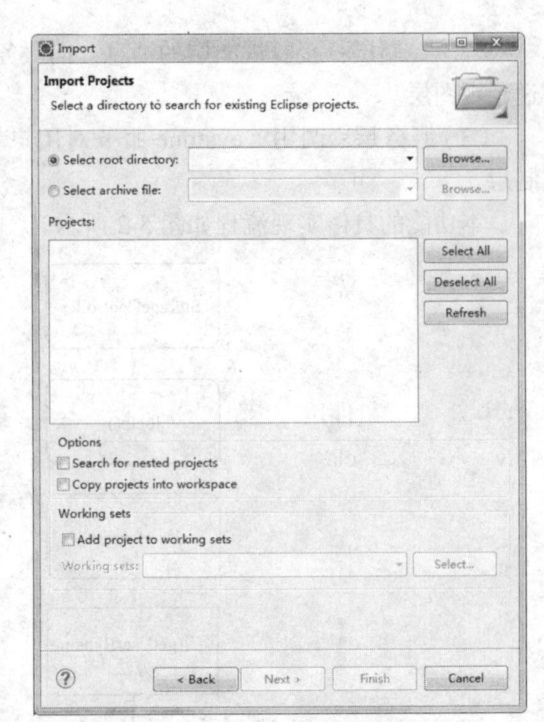


图 5-4　Eclipse 导入工程路径

2. 实现控制层

在“/src/com/xiandian/cloud/storage/Web/StorageController.Java”增加如下代码：

```
/**
     * 登录
     * @param request:请求
     * @param username:用户名
     * @param password:密码
     */
    @RequestMapping("/login")
    @ResponseBody
    public Object login(HttpServletRequest request,String username, String
password) {
        //通过 username 来获取数据库中对应 user 的信息
User user = userService.getUserByname(username);
        if (user == null) {
            return new MessageBean(false,Constants.ERROR_1);
        }
        //这段代码用于用户密码解密使用，如数据库中数据未加密可去掉此段密码
        String inputstr = QEncodeUtil.aesDecrypt(user.getPassword(), QEncodeUtil.
ydy);
        if (!inputstr.equals(password)) {
```

```
            return new MessageBean(false,Constants.ERROR_2);
        }
        setSessionUser(request, user);
        return new MessageBean(true,Constants.SUCCESS_1);
    }
```

**3. 实现视图层**

（1）在/WebRoot/WEB-INF 文件夹下新建文件 index.html。登录界面采用 Bootstrap panels 面板及 Bootstrap form 表单来实现。

（2）面板组件用于把 DOM 组件插入到一个盒子中。创建一个基本的面板，只需要向<div>元素添加 class .panel 和 class .panel-default 即可，如下所示。

```
<div class="panel panel-default">
    <div class="panel-body">
      <!--登录 form 表单代码区域-->
    </div>
</div>
```

（3）在本实例中可以看到，除了面板主体部分以外还有一个面板标题，可以通过以下两种方式来添加面板标题：

① 使用“.panel-heading class”向面板添加标题容器。

② 使用带有“.panel-title class”的<h1>-<h6>添加预定义样式的标题。

```
<div class="panel panel-default">
 <div class="panel-heading">
       <h3 class="panel-title">
           登录云网盘
       </h3>
    </div>
    <div class="panel-body">
      <!--登录 form 表单代码区域-->
    </div>
</div>
```

这样就实现了一个简单的登录面板，如图 5-5 所示。

图 5-5　登录面板

接下来开始在这个面板的 body 区域加入 form 表单，以支持用户输入用户名和密码等信息。

（4）登录表单实现：本案例采用垂直表单实现用户登录界面，步骤如下。

① 向父<form>元素添加 “role="form"”。

② 把标签和控件放在一个带有“class .form-group”的<div>中。这是获取最佳间距所必需的。

③ 向所有的文本元素<input>、<textarea>和<select>添加“class ="form-control"”。

```
<form role="form">
    <div class="form-group">
        <label for="name">账号</label>
        <input type="text" class="form-control" id="username" placeholder="请
输入名称">
        <div class="help-block">请输入用户昵称</div>
    </div>
    <div class="form-group">
        <label for="password">密码</label>
    <input type="password" class="form-control" id="username" placeholder="请
输入密码">
    </div>
    <div class="form-group">
    <button  style="width:100%"  id="_login"  class="btn  btn-primary"> 登 录
</button>
    </div>
</form>
```

④ 最后将 form 表单嵌入到 panel-body 区域中，登录界面如图 5-6 所示。

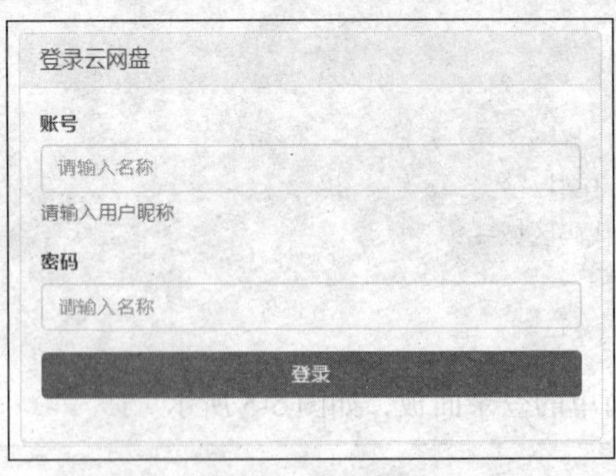


图 5-6　登录界面

```
<div class="panel panel-default">
<div class="panel-heading">
        <h3 class="panel-title">
            登录云网盘
        </h3>
```

```
    </div>
    <div class="panel-body">
     <form role="form">
     <div class="form-group">
         <label for="name">账号</label>
         <input type="text" class="form-control" id="username" placeholder="请
输入名称">
         <div class="help-block">请输入用户昵称</div>
     </div>
     <div class="form-group">
         <label for="password">密码</label>
     <input type="password" class="form-control" id="password" placeholder="请
输入密码">
     </div>
     <div class="form-group">
     <button  style="width:100%"  id="_login"  class="btn  btn-primary"> 登 录
</button>
     </div>
</form>
     </div>
</div>
```

4. 实现前后台数据交互

在上一节内容中，已经完成了登录界面的编写，接下来就要与后端进行数据交互，本案例中采用 Ajax 与 JSON 实现数据交互。

在界面代码 index.html 中，找到 id= "_login" 的 div 模块，为其添加单击事件 login() 方法：

```
<script type="text/javascript">
        //回车登录
        function keyLogin() {
            if (event.keyCode == 13)
                document.getElementById("_login").click();
        }
        //登录
        function login() {
            var username = $("#username").val();    //获取用户输入的用户名
            var password = $("#password").val();    //获取用户输入的密码
```

```
        var data = {                              //传输到后台的json数据
            username : username,
            password : password
        };
        $.ajax({
            url : "login.action",             //与后台实现交互的登录接口
            type : "post",                    //通过post方式传输到后台
            data : data,                      //将data通过ajax传输到后台
            success : function(s) {           //传输成功后接收后台返回数据
                if (s.success) {
                    location.href = "home.action";
                } else {
                    alert(s.msg);
                }
            }
        });
    }
</script>
```

5. 功能验证测试

将项目部署到 Tomcat 中并发布，访问 http://localhost:8080，显示界面如图 5-7 所示，填入已经注册好的用户名和密码，进行登录验证。

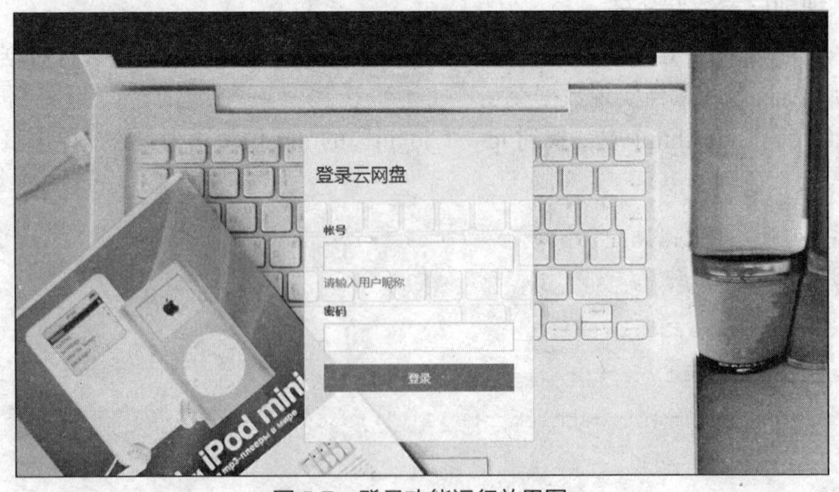


图 5-7 登录功能运行效果图

登录功能的测试场景见表 5-2。

表 5-2　登录功能测试场景

| 编号 | 测试场景 | 输入参数 | 预期结果 |
| --- | --- | --- | --- |
| 1 | 用户名称和密码都正确 | Username：gw001<br>Password：123456 | 登录成功，进入所有文件界面 |
| 2 | 用户名称正确，密码错误 | Username：gw001<br>Password：000000 | 登录不成功，提示用户名或密码错误，请重新输入 |
| 3 | 用户名称错误，密码可以任意 | Username：gw000<br>Password：123456 | 登录不成功，提示用户名或密码错误，请重新输入 |
| 4 | 用户名称为空，密码任意 | Username：<br>Password：123456 | 提示用户名称不能为空，请重新输入 |
| 5 | 用户名称任意，密码为空 | Username：123456<br>Password： | 提示用户密码不能为空，请重新输入 |

## 任务 5.2　开发注册功能

### 5.2.1　相关知识

本功能实现的技术原理如下。

（1）视图层：设计如图 5-8 所示的注册界面。

图 5-8　网盘注册界面

（2）控制层：接收视图层的消息，向服务层发送注册信息，并将服务层返回的结果发送到视图层。

（3）服务层：调用 Keystone 服务对用户名和密码进行认证，并将认证结果返回控制层。

本功能的具体实现流程如图 5-9 所示。

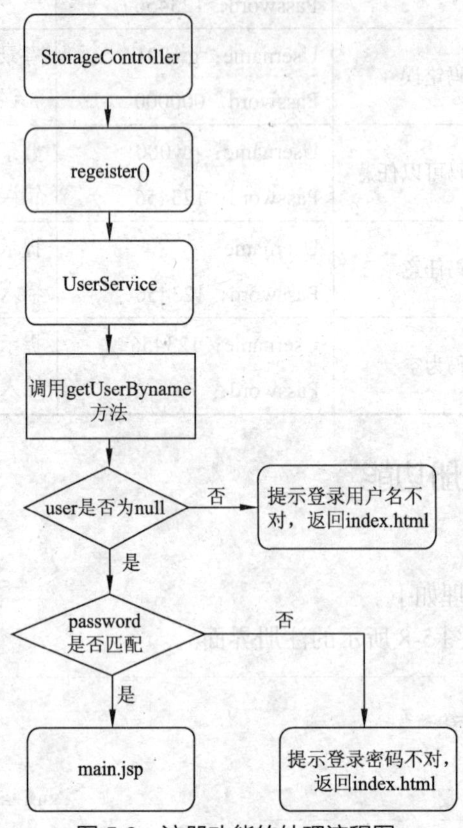


图 5-9　注册功能的处理流程图

### 5.2.2　实现步骤

#### 1．实现控制层

在“src/com/xiandian/cloud/storage/web/UserController.java”中增加如下代码：

```
/**
     * 注册
     * @param request:请求
     * @param username:用户名
     * @param password:密码
     */
    @RequestMapping("/regeister")
    @ResponseBody
    public Object regeister(HttpServletRequest request,String username,String
password) {
```

```
        User user = userService.getUserByname(username);
        if (user != null) {
            return new MessageBean(false,Constants.ERROR_3);
        }
        user = userService.save(username,password);
        setSessionUser(request, user);
        //新建用户，创建租户、用户、容器、回收站
        storageService.createUser(username, password);
        return new MessageBean(true,Constants.SUCCESS_2);
    }
```

2. 实现服务层

（1）在“src/com/xiandian/cloud/storage/service/UserService.java”中增加如下代码：

```
/**
     * 创建用户
     * @param username:用户名
     * @param password:密码
     */
    public void createUser(String username, String password) {
        try {
            String OPENSTACK_IP = SwiftUtilTools.getConfig().getProperty("AUTHURL");
            String OPENSTACK_USER_NAME = SwiftUtilTools.getConfig().getProperty
("USERNAME");
            String OPENSTACK_USER_PW = SwiftUtilTools.getConfig().getProperty
("PASSWORD");
            String OPENSTACK_TENANE_NAME = SwiftUtilTools.getConfig().getProperty
("USERNAME");
            String keystoneAuthUrl = "http://" + OPENSTACK_IP + ":35357/v2.0";
            Keystone keystone = new Keystone(keystoneAuthUrl);
            // access with unscoped token
            Authenticate abc = keystone.tokens().authenticate(new UsernamePassword
(OPENSTACK_USER_NAME,OPENSTACK_USER_PW));
            Access access = abc.execute();
            access = keystone.tokens().authenticate(new TokenAuthentication
(access.getToken().getId())).withTenantName(OPENSTACK_TENANE_NAME).execute();
            Tenant tenant = new Tenant();
            tenant.setName(username);
            tenant.setDescription(username);
            tenant.setEnabled(true);
```

```
            // Get the adminURL client and use the token got above
            keystone = new Keystone(keystoneAuthUrl);
            keystone.token(access.getToken().getId());
            tenant = keystone.tenants().create(tenant).execute();
            User user = new User();
            user.setUsername(username);
            user.setPassword(password);
            user.setName(username);
            user.setEnabled(Boolean.TRUE);
            user.setTenantId(tenant.getId());
            user = keystone.users().create(user).execute();
            keystone.tenants().addUser(tenant.getId(),
user.getId(),SwiftUtilTools.getConfig().getProperty("ROLEID")).execute();
            String inputstr = QEncodeUtil.aesEncrypt(password, QEncodeUtil.ydy);
            Account account = SwiftUtilTools.getAccount(username, inputstr);
            SwiftStoreImpl Swiftdfs = new SwiftStoreImpl(account);
            Swiftdfs.createContainer(username);
            Swiftdfs.createContainer(Constants.GARBAGE_PREFIX + username);
        } catch (Exception e) {
            e.printStackTrace();
        }
    }
```

（2）在“src/com/xiandian/cloud/storage/sh/SwiftStoreImpl.java”中增加如下代码：

```
private Container Container;
    private Account account = null;
    public SwiftStoreImpl() {
    }
    /**
     * 为了少改动代码，增加传入 Account 参数的构建器
     *
     * @param account
     */
    public SwiftStoreImpl(Account account) {
        this.account = account;
    }
    /**
     * 描述：创建容器，根据根路径
     *
```

```
     * @param rootPath
     *            根路径
     * @return boolean
     */
    public boolean createContainer(String rootPath) {
        Container Containerc = account.getContainer(rootPath);
        if (Containerc.exists()) {
            return false;
        } else {
            // logger(rootPath+"执行 createContainer 方法，创建容器操作");
            return createContainer(Containerc);
        }
    }
    private boolean createContainer(Container Container) {
        Container cona = Container.create();
        return cona.exists();
    }
```

### 3. 实现视图层

在/WebRoot/WEB-INF/jsp 文件夹下新建 regeister.jsp。注册界面采用 Bootstrap form 水平表单来实现。

水平表单与其他表单不仅标记的数量不同，而且表单的呈现形式也不同。如需创建一个水平布局的表单，需按下面的几个步骤进行。

（1）向父<form>元素添加“class .form-horizontal”。

（2）把标签和控件放在一个带有“class .form-group” 的<div>中。

（3）向标签添加“class .control-label”。

```
<form class="form-horizontal" role="form">
<div class="form-group">
  <label for="username" class="col-sm-2 control-label">
    <span style="color:red;">*</span>
     用户名:
    </label>
   <div class="col-sm-10">
    <input class="form-control" id="username" type="text" placeholder="例如:
zhangsan" />
    </div>
   </div>
   <div class="form-group">
    <label for="password" class="col-sm-2 control-label">
```

```
        <span style="color:red;">*</span>
        密码:
      </label>
        <div class="col-sm-10">
          <input class="form-control" id="password" type="password" />
          </div>
      </div>
  <div class=" form-group" >
    <div class="col-sm-12">
      <button id="regesiter" type="submit" class="btn btn-primary btn-large">
注册</button>
    </div>
  </div>
</form>
```

4. 实现前后台数据交互

在界面代码中找到“id="regesiter"”的 button 模块，为其添加单击事件“regeister()”：

```
<script>
    //注册
        function regeister() {
            var password = $("#password").val();    //获取用户密码
            var username = $("#username").val();    //获取用户名
            if(password==""){                         //验证密码是否为空
                alert("密码不能为空!");
                return false;
            }
            if(username==""){                         //验证用户名是否为空
                alert("用户名不能为空!");
                return false;
            }
            var data = {                              //传输到后台的 json 数据
                username : username,
                password : password
            };
            $.ajax({
                url : "regeister.action",             //与后台实现交互的注册接口
                type : "post",                        //通过 post 方式传输到后台
                data : data,                          //将 data 通过 ajax 传输到后台
                success : function(s) {               //传输成功后接收后台返回数据
```

```
                if (s.success) {
                    location.href = "home.action"; //成功接收跳转进入网盘
                }
                    else {
                        alert(s.msg);
                    }
                }
            });
        }
</script>
```

5. 功能验证测试

将项目部署到 Tomcat 中并发布，项目运行成功后，登录进入，打开注册界面，填入注册的用户名和密码，单击“注册”按钮，如图 5-10 所示，注册完成后进入网盘主界面，显示所有文件界面。

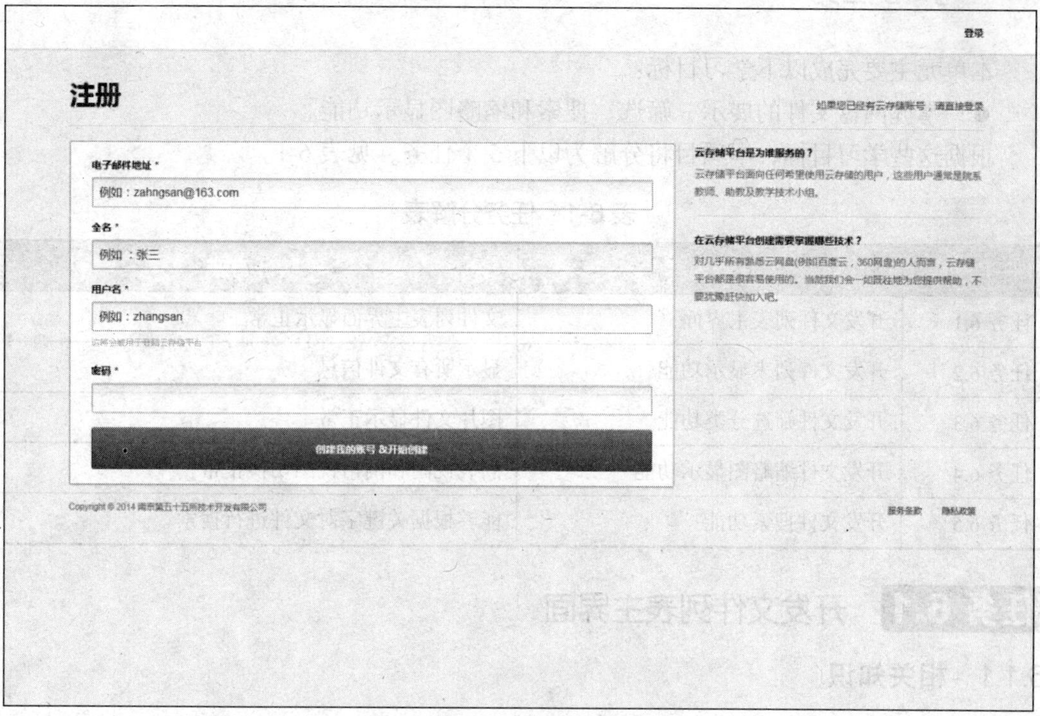

图 5-10　注册功能运行效果图

# 项目 6 开发文件列表模块

## 单元介绍

本单元读者需要掌握网盘文件列表模块中功能的实现，包括前后台代码的流程，并对其功能进行验证测试。

## 学习任务

本单元主要完成以下学习目标：

- 掌握网盘文件的展示、筛选、搜索和缩略图显示功能。

根据这些学习目标，本项目将分解为以下 5 个任务，见表 6-1。

表 6-1 任务分解表

| 任务序号 | 任务内容 | 验证方式 |
|---|---|---|
| 任务 6.1 | 开发文件列表主界面 | 文件列表主界面显示正常 |
| 任务 6.2 | 开发文件列表显示功能 | 显示所有文件信息 |
| 任务 6.3 | 开发文件筛选分类功能 | 图片文件显示正常 |
| 任务 6.4 | 开发文件缩略图显示功能 | 列表视图和网格视图切换正常 |
| 任务 6.5 | 开发文件搜索功能 | 能够根据关键字对文件进行搜索 |

## 任务 6.1 开发文件列表主界面

### 6.1.1 相关知识

文件列表模块的界面如图 6-1 所示。

界面显示所有文件夹和文件，文件名称按字母排序排列，先显示文件夹，再显示文件。

图 6-1 文件列表主界面

### 6.1.2 实现步骤

在正式编写 HTML 代码之前，首先分析整个网页结构。整个网页由头部、主体内容两块区域构成，而主体内容部分又可以拆分为导航区域和文件内容区域。页面整体框架如下：

```
<html lang="zh-cn">
<head>
<meta charset="utf-8">
<meta http-equiv="X-UA-Compatible" content="IE=edge">
<meta name="viewport" content="width=device-width, initial-scale=1">
<meta name="description" content="">
<meta name="author" content="">
<title>先电云存储</title>
</head>
<body>
<!--头部区域代码-->
 ...
<div class="main">
<!--左侧导航区域代码-->
...
<!--右侧内容区域代码-->
...
</div>
```

```
</body>
</html>
```

1. 引入 CSS 和基础 JavaScript 文件

在“<head></head>”标签区域，引入相关 CSS 样式文件。

```
<link href="assets/stylesheets/bootstrap.min.css" rel="stylesheet"type="text/css" />
<link href="assets/stylesheets/style.css" rel="stylesheet" type="text/css" />
<link href="assets/stylesheets/disk.css" rel="stylesheet" type="text/css" />
<link href="assets/stylesheets/font-awesome.css" rel="stylesheet" type="text/css" />
<link href="assets/stylesheets/zTreeStyle.css" rel="stylesheet" type="text/css" />
```

在“<body></body>”标签区域，引入相关 js 文件。

```
<script src="assets/javascripts/jquery.min.js"></script>
<script src="assets/javascripts/bootstrap.min.js"></script>
<script src="assets/javascripts/fineuploader.js"></script>
<script src="assets/javascripts/checkbox.js"></script>
<script src="assets/javascripts/common.js"></script>
```

2. 实现头部区域

引用 CSS 样式文件及 js 文件后，开始编写头部区域。头部区域代码由两块内容组成：topbar 部分和 Frame Header 部分。

（1）topbar 部分的实现。

```
<div class="topbar">
 <div class="content">
 <div class="top_left">
 <a href="javascript:void(0)" class="topbar_menu title" draggable="false">
 <i class="fa fa-cloud"></i>Cloud
 </a>
 <a class="topbar_menu this" target="_self" draggable="false">
 <i class="font-icon menu-explorer"></i>文件管理
 </a>
 </div>
 <div class="top_right">
 <div class="menu_group">
 <a href="#" id="topbar_user" data-toggle="dropdown"
    class="topbar_menu" draggable="false">
 <i class="font-icon icon-user"></i>管理员 <b class="caret"></b></a>
 <ul class="dropdown-menu menu-topbar_user pull-right animated menuShow"
role="menu" aria-labelledby="topbar_user">
```

```
<li>
<a href="#" draggable="false">
<i class="font-icon fa fa-sign-out"></i>个人信息</a>
</li>
<li>
<a href="#" draggable="false">
<i class="font-icon fa fa-sign-out"></i>退出</a></li></ul>
</div>
</div>
<div style="clear: both"></div>
</div>
</div>
```

（2）Frame Header 部分的实现。

```
<div class="frame-header">
<div class="header-content">
<div class="header-left">
<div class="btn-group btn-group-sm">
<button onclick="javascript:history.go(-1);" class="btn btn-default"
id="history_back" title="后退" type="button">
<i class="font-icon fa fa-arrow-left"></i>
</button>
<button onclick="javascript:history.go(1);" class="btn btn-default"
   id="history_next" title="前进" type="button">
<i class="font-icon fa fa-arrow-right"></i></button>
<button onclick="javascript:location.reload();" class="btn btn-default"
id="refresh" title="强制刷新"  type="button">
<i class="font-icon fa fa-refresh"></i>
</button>
</div>
</div>
<div class="header-middle">
<a href="#">
<button class="btn btn-default" id="home" title="我的文档">
<i class="font-icon fa fa-home"></i></button>
</a>
<div id="yarnball">
<ul class="yarnball">
<li class="yarnlet first">
```

```
<a title="" href="#">
<span class="left-yarn"></span>
<span class="address_ico groupSelf"></span>
<span class="title_name">全部</span>
</a></li></ul></div>
<div id="yarnball_input" style="display: none;">
<input type="text" name="path" value="" class="path" id="path">
</div></div>
<div class="header-right">
<input type="text" name="seach" id="navbarInput-01">
<a class="btn btn-default" id="searchfiles" title="搜索" type="button">
<i class="font-icon fa fa-search"></i>
</a></div></div></div>
```

3. 实现导航区域

编写好头部区域后，开始左侧导航栏的编写。导航区域主要由 8 个导航体组成。

```
<div class="main-left" style="z-index: 1;">
 <div class="main-left" style="z-index: 1;">
 <div class="list-group category" style="margin: 10px;">
 <div id="ahref">
 <a class="list-group-item active" id="r_active">
  <span class="glyphicon glyphicon-home left-icon"></span>
     全部文件 </a>
 <a class="list-group-item " id="type">
  <span class="glyphicon glyphicon-file left-icon"></span>
    我的文档</a>
 <a class="list-group-item ">
 <span class="glyphicon glyphicon-picture left-icon"></span>
    我的图片</a>
 <a class="list-group-item " id="type">
 <span class="glyphicon glyphicon-film left-icon"></span>
   我的视频</a>
 <a class="list-group-item " id="type">
 <span class="glyphicon glyphicon-music left-icon"></span>
   我的音乐</a>
 <a class="list-group-item" id="type">
 <span class="glyphicon glyphicon-inbox left-icon"></span>
   其他</a>
 <a class="list-group-item" id="type">
```

```
<span class="glyphicon glyphicon-share left-icon"></span>
  我的分享</a>
<a class="list-group-item" >
<span class="glyphicon glyphicon-trash left-icon"></span>
  垃圾箱</a></div></div>
<div class="main-left-use">
<div class="progress progress-u progress-xs">
<div class="progress-bar progress-bar-blue" id="progress-bar"
role="progressbar" aria-valuemin="0" aria-valuemax="100"></div>
</div>
<h3 class="heading-xs">容量:
<span id="totle"></span>
<span class="capacity pull-right">
<a href="javascript:void(0)">扩容</a>
</span></h3></div></div></div>
```

4. 实现文件内容区域

文件内容区域又可细分为两块，一块是按钮组，还有一块是文件列表。在相关知识中本书已经介绍过 Bootstrap 相关按钮组件，以及表格组件的知识，下面来实际应用这些相关组件。

文件内容区域由按钮组和文件列表组成，结构如下。

```
<div class="main-right">
  <!--按钮组-->
    <div class="col-md-12" style="padding: 0"></div>
  <!--文件列表-->
     <div class="col-md-12" style="padding: 0"></div>
  </div>
```

按钮组的实现：按钮组由上传、新建文件夹、删除、下载、重命名、复制、移动、数据分析等按钮组成。按钮组用于对网盘文件的操作。

```
<div class="tools">
 <div class="tools-left">
 <div class="btn-group btn-group-sm kod_path_tool fl-left">
 <div id="result-uploader" class="right upload-filemain fl pull-left"></div>
 <button class="btn btn-default" type="button" id="newdir">
 <i class="font-icon fa  fa-folder-open-o"></i>新建文件夹
 </button>
 <button class="btn btn-default" style="display: none;"type="button"id="delete">
 <i class="font-icon glyphicon glyphicon-trash" style="color: #888"></i>删除
```

```
</button>
<a class="btn btn-default" type="button" style="display: none;"  id="download">
<i class="font-icon glyphicon glyphicon-download-alt" style="color: #888"></i>
下载
</a>
<button class="btn btn-default" style="display: none;" type="button" id="rename">
<i class="font-icon glyphicon glyphicon-pencil" style="color: #888"></i>
重命名
</button>
<a class='btn btn-default' style="display: none;" id="download">
<span lass="glyphicon glyphicon-download-alt"><span>下载文件</a>
<a class='btn btn-default' style="display: none;" id="copy">
<span class="glyphicon glyphicon-file"></span>复制</a>
<a class='btn btn-default' style="display: none;" id="move">
<span class="glyphicon glyphicon-transfer"></span>移动</a>
<a class='btn btn-default' style="display: none;" id="rename">
<span class="glyphicon glyphicon-pencil"></span>重命名</a>
<a class='btn btn-default' style="display: none" id="Dataanalysis"></a>
<ul class="pull-left">
<li class="dropdown">
<a class="dropdown-toggle btn btn-default" data-toggle="dropdown"
  id="selectcolumn" style="display: none" href="javascript:void(0);">
 数据分析</a>
<ul class="dropdown-menu cloudanys" style="margin-top: 4px !important;">
<li><a href="javascript:void(0);">词云分析</a></li>
<li class="divider"></li>
<li><a href="javascript:void(0);">柱状分析</a></li>
<li class="divider"></li>
<li><a href="javascript:void(0);">气泡分析</a></li>
</ul></li></ul></div>
<div class="clearfix"></div>
</div></div>
```

文件列表：文件列表区域主要使用 Bootstrap 表格布局。

```
<table class="table mb-0">
 <thead>
 <tr>
 <th class="table-checkbox" style="position: relative; left: 13px;">
 <label class="checkbox checkbox-position" for="checkbox1">
```

```
<span class="icons icon-span">
<span class="first-icon fui-checkbox-unchecked"></span>
<span class="second-icon fui-checkbox-checked"></span>
</span>
<input name="chkAll" type="checkbox" id="operAll" value="checkbox"
      class="main-check"/></label></th>
<th class="mainfile-name">文件名</th>
<th class="hide table-fileposition mainfile-position">文件位置</th>
<th class="mainfile-size">大小</th>
<th class="mainfile-size" id="share_th">分享</th>
<th class="mainchange-date">修改日期</th>
</tr>
</thead>
<tbody id="tab">
<tr><td>
<label class="checkbox table-checkboxposition" for="checkbox1">
<span class="icons main-icons">
<span class="first-icon fui-checkbox-unchecked"></span>
<span class="second-icon fui-checkbox-checked"></span>
</span> <input type="checkbox" name='check' class="main-tabinput">
</label></td><td>
<span style="display: block">cloud.doc</span>
<div class="edit-name" style="display: none;">
<input class="box"  type="text" value="cloud.doc">
<a class="sure" href="javascript:void(0);">
<span class="glyphicon glyphicon-ok"></span></a>
<a class="cancel ml-10" href="javascript:void(0);">
<span class="glyphicon glyphicon-remove"></span></a>
</div></td>
<td class="hide table-fileposition table-path">cloudskill/</td>
<td>23 Bytes</td>
<td id="share_td"><a type="button">
<img src="assets/images/share2.png" width=20px height=20px />
</a>
</td>
<td>2017-01-11</td></tr></tbody>
</table>
```

## 任务 6.2 开发文件列表显示功能

用户登录完成后，主界面显示所有文件的列表。

### 6.2.1 相关知识

本功能实现的技术原理如下。

（1）视图层：展示所有文件信息。

（2）控制层：接收视图层的消息，向服务层发送文件列表参数，并将服务层返回的结果发送到视图层。

（3）服务层：获取当前账号的所有云存储对象，将获取到的云存储对象转换为本地模拟文件系统。

本功能的具体实现流程如图 6-2 所示。

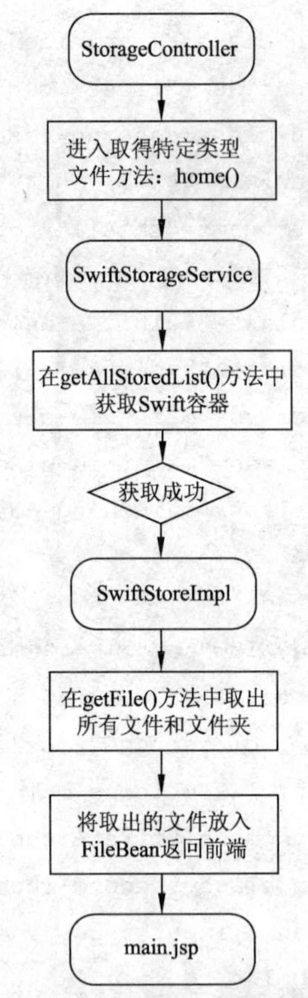


图 6-2 文件列表显示功能的处理流程

### 6.2.2 实现步骤

#### 1. 项目导入

在 Eclipse 中导入工程“project61”。

#### 2. 实现服务层

（1）在 src/com/xiandian/cloud/storage/sh/bean 目录下新建一个名为 FileBean 的类，该类定义了文件的一些属性，包括文件名称、文件的大小、文件路径、文件修改时间、是否为目录、是否有子文件夹等，这些信息将会被传到前台页面显示，具体代码如下：

```
import com.xiandian.cloud.storage.sh.util.UtilTools;
/**
 * 与前台交互数据的 bean
 * @author 云计算应用与开发项目组
 * @since  V1.4
 *
 */
public class FileBean {
    //是否是目录
    private boolean isdirectory;
    //是否有子文件夹
    private boolean haschild;
    //文件名称
    private String name;
    //文件路径
    private String path;
```

```
//文件大小
private String length;
//最后修改时间
private String lastmodified;
//如果是文件，存放文件的临时路径
private String filepath;
public boolean isHaschild() {
    return haschild;
}
public void setHaschild(boolean haschild) {
    this.haschild = haschild;
}
public boolean isIsdirectory() {
    return isdirectory;
}
public void setIsdirectory(boolean isdirectory) {
    this.isdirectory = isdirectory;
}
public String getFilepath() {
    return filepath;
}
public void setFilepath(String filepath) {
    this.filepath = filepath;
}
public String getLength() {
    return length;
}
public void setLength(long length) {
    this.length = UtilTools.convertFileSize(length);
}
public String getLastmodified() {
    return lastmodified;
}
public void setLastmodified(String lastmodified) {
    this.lastmodified =lastmodified;
}
public String getName() {
    return name;
}
```

```
    public void setName(String name) {
        this.name = name;
    }
    public String getPath() {
        return path;
    }
    public void setPath(String path) {
        this.path = path;
    }
}
```

（2）在"src/com/xiandian/cloud/storage/sh/SwiftStoreImpl.java"类中增加如下代码：

```
/**
     * 获取文件类表
     *
     * @param rpath 当前目录
     * @return List
     */
    public List getFile(String rpath) {
        Map mappath = getSplitPath(rpath);
        String path = mappath.get("path").toString();
        Container container = account.getContainer(mappath.get("rootPath")
                .toString());
        Collection<DirectoryOrObject> objects;
        if (path != null && path.length() > 1) {
            Directory directory = new Directory(path, '/');
            objects = listDirectory(container, directory);
        } else {
            objects = listDirectory(container);
        }
        List list = new ArrayList();
        for (DirectoryOrObject c : objects) {
            FileBean fb = new FileBean();
            fb.setName(c.getBareName());
            fb.setIsdirectory(c.isDirectory());
            if (c.isDirectory()) {
                fb.setPath(c.getAsDirectory().getName());
                list.add(fb);
            } else {
```

```
                StoredObject t = c.getAsObject();
                String tp = t.getPath();
                String ttp = SwiftUtilTools.replaceStr(tp);
                if (!ttp.endsWith("/")) {
                    fb.setPath(t.getName());
                    fb.setLength(t.getContentLength());
                    fb.setLastmodified(t.getLastModified());
                    list.add(fb);
                }
            }
        }
        return list;
    }
    private Map<String, String> getSplitPath(String path) {
        String[] strPath = path.split("/");
        String rootPath = "";
        path = "";
        for (int i = 0; i < strPath.length; i++) {
            if (i == 0) {
                rootPath = strPath[0];
            } else {
                if (i == strPath.length - 1) {
                    if (strPath[i].contains(".")) {
                        path += strPath[i];
                    } else {
                        path += strPath[i] + "/";
                    }
                } else {
                    path += strPath[i] + "/";
                }
            }
        }
        Map map = new HashMap();
        map.put("rootPath", rootPath);
        map.put("path", path);
        return map;
    }
    private Collection listDirectory(Container container, Directory directory)
{
```

```
        return container.listDirectory(directory);
    }
    private Collection listDirectory(Container container) {
        return container.listDirectory();
    }
```

（3）在 com/xiandian/cloud/storage/util 目录下新建一个名为 ComparatorFile 的类，该类对文件和文件夹进行排序，由界面调用，在文件列表中，先显示文件夹再显示文件。具体代码如下：

```
package com.xiandian.cloud.storage.util;
import java.util.Comparator;
import com.xiandian.cloud.storage.sh.bean.FileBean;
/**
 * 文件夹与文件排序的实现类
 *
 */
public class ComparatorFile implements Comparator<FileBean> {
    public ComparatorFile() {
    }
    public int compare(FileBean f1, FileBean f2) {
        boolean f1dir = f1.isIsdirectory();
        boolean f2dir = f2.isIsdirectory();
        if (f1dir && f2dir) {
            return 0;
        } else if (!f1dir && f2dir) {
            return 1;
        } else if (f1dir && !f2dir) {
            return -1;
        } else if (!f1dir && !f2dir) {
            return 0;
        }
        return 0;
    }
}
```

（4）在“src/com/xiandian/cloud/storage/service/SwiftStorageService.java”类中增加如下代码：

```
/**
    * 取得路径下的所有文件
    */
```

```
    public List getAllStoredList(String username, String password, String path)
{
        Account account = SwiftUtilTools.getAccount(username, password);
        SwiftStoreImpl Swiftdfs = new SwiftStoreImpl(account);
        String up = null;
        up = username + "/" + path;
        List list = Swiftdfs.getFile(up);
        ComparatorFile comparator = new ComparatorFile();
        if (!list.isEmpty()) {
            synchronized (list) {
                Collections.sort(list, comparator);
            }

        }
        return list;
}
```

3. 实现控制层

在“/src/com/xiandian/cloud/storage/Web/StorageController.java”类中增加如下代码：

```
/**
    * 取得主页的文件
    *
    * @param request
    * @param response
    * @param path
    * @return
    */
   @RequestMapping("/home")
   public ModelAndView home(HttpServletRequest request,
           HttpServletResponse response, String path) {
       ModelAndView view = new ModelAndView();
       User user = getSessionUser(request);
       path = UtilTools.converStr(path);
       List list = storageService.getAllStoredList(user.getUsername(),
               user.getPassword(), path);
       view.addObject("path", path);
       view.addObject("search", "false");
       view.addObject("list", list);
       view.setViewName("/main");
```

```
        view.addObject("type", 0);
        return view;
    }
```

4．实现前后台数据交互

根据控制层“StorageController”类展示所有文件方法 home()，可以看到后台列表数据是跟随页面一起返回的，因此在页面中本书采用 JSTL 中“<c:forEach>”及“<c:when>”方法循环获取网盘中的目录和文件消息。

```
<tbody id="tab">
 <c:forEach var="fb" items="${list}">
 <tr>
  <td><label lass="checkbox table-checkboxposition" for="checkbox1">
  <span class="icons main-icons">
  <span class="first-icon fui-checkbox-unchecked"></span>
  <span class="second-icon fui-checkbox-checked"></span>
  </span>
 <input type="checkbox" name='check'class="main-tabinput"onclick="show()">
 </label>
 </td>
 <td><span style="display: block">
 <c:choose>
 <c:when test="${fb.isdirectory == true}">
 <a href="home.action?path=${fb.path}">
 </c:when>
 <c:otherwise>
 <a href="javascript:void(0);" onclick="showread('${fb.path}');">
 </c:otherwise>
 </c:choose>
 <c:if test="${fb.isdirectory == true}">
 <img src="assets/images/0.png" class="objimg">
 <input name="objimg" type="text" style="display: none" value="${fb.name}">
 </c:if> </a> ${fb.name}</span>
 <div class="edit-name" style="display: none;">
 <input class="box" type="text" value="${fb.name}">
 <a class="sure" href="javascript:void(0);" onclick="sure()">
 <span class="glyphicon glyphicon-ok"></span></a>
 <a class="cancel ml-10" onclick="cancel()" href="javascript:void(0);">
 <span class="glyphicon glyphicon-remove"></span>
 </a></div></td>
```

```
<td class="hide table-fileposition table-path">${fb.path}</td>
<td>${fb.length}</td>
<td id="share_td"><c:if test="${fb.length != null}">
<a type="button"><img src="assets/images/share2.png"/></a>
</c:if>
</td>
<td>${fb.lastmodified }</td>
</tr>
</c:forEach>
</tbody>
```

5. 功能验证测试

将项目部署到 Tomcat 中并发布，项目运行成功后，登录进入，单击侧边栏的“全部文件”按钮，显示所有文件，运行效果如图 6-3 所示。

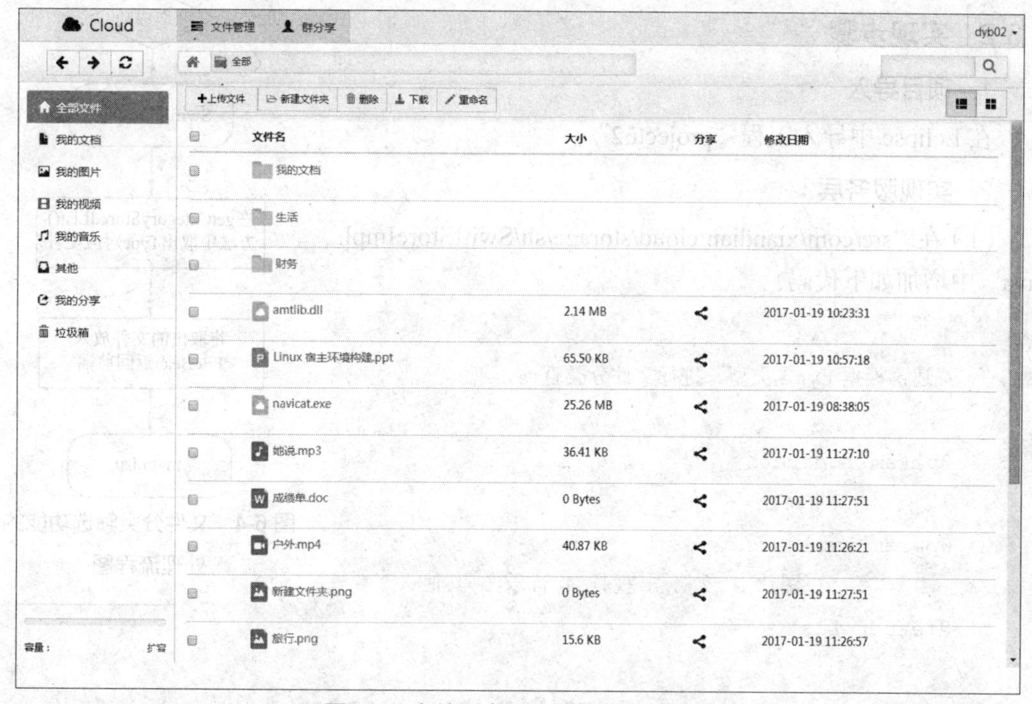


图 6-3　文件列表显示功能运行效果图

本功能的测试场景见表 6-2。

表 6-2　文件列表显示功能测试场景

| 编号 | 测试场景 | 输入参数 | 预期结果 |
| --- | --- | --- | --- |
| 1 | 登录成功后 | 单击“登录”按钮 | 登录成功，进入显示所有文件界面 |
| 2 | 登录成功后 | 单击“全部文件”按钮 | 进入显示所有文件界面 |

## 任务 6.3 开发文件筛选分类功能

单击左侧边栏的不同模块可以对文件进行分类筛选显示。例如，单击“我的文档”按钮显示当前用户的所有文档文件。

### 6.3.1 相关知识

本功能实现的技术原理如下。

（1）视图层：分类展示图片。

（2）控制层：接收视图层的消息，向服务层发送文件筛选参数，并将服务层返回的结果发送到视图层。

（3）服务层：获取当前账号的所有图片对象，将图片对象封装到集合中返回给控制层。

本功能的具体实现流程如图 6-4 所示。

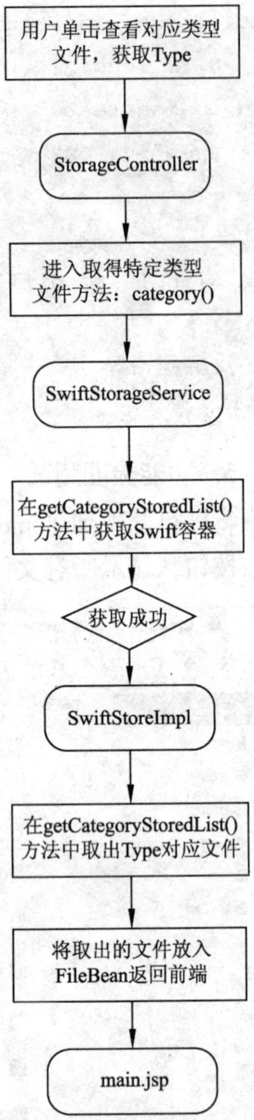

用户单击查看对应类型文件，获取Type

StorageController

进入取得特定类型文件方法：category()

SwiftStorageService

在getCategoryStoredList()方法中获取Swift容器

获取成功

SwiftStoreImpl

在getCategoryStoredList()方法中取出Type对应文件

将取出的文件放入FileBean返回前端

main.jsp

图 6-4 文件分类筛选功能的处理流程图

### 6.3.2 实现步骤

#### 1. 项目导入

在 Eclipse 中导入工程“project62”。

#### 2. 实现服务层

（1）在“src/com/xiandian/cloud/storage/sh/SwiftStoreImpl.java”中增加如下代码：

```
/**
    * 描述：根据 type 类型，进行文件分类查询
    *
    * @param rootPath
    *            根路径
    * @param type
    *            1 图片 2 文档 3 视频 4 音乐 5 其他
    * @return List
    */
   public List getCategoryStoredList(String rootPath, int type) {
       Container container = account.getContainer(rootPath);
       Collection<StoredObject> objects = list(container);
       List<StoredObject> objS = new ArrayList();
       for (StoredObject c : objects) {
           if(c.getAsObject().getContentLength() == 0){ // isDir 判断无效
               continue;
           }
           String name = c.getName();
```

```
            if (type == 1) {
                if (SwiftUtilTools.isImage(name)) {
                    objJS.add(c);
                }
            } else if (type == 2) {
                if (SwiftUtilTools.isDoc(name)) {
                    objJS.add(c);
                }
            } else if (type == 3) {
                if (SwiftUtilTools.isMediea(name)) {
                    objJS.add(c);
                }
            } else if (type == 4) {
                if (SwiftUtilTools.isMp3(name)) {
                    objJS.add(c);
                }
            } else if (type == 5) {
                if (!SwiftUtilTools.isImage(name) && !SwiftUtilTools.isDoc(name)
                        && !SwiftUtilTools.isMediea(name) && !SwiftUtilTools.
isMp3(name)) {
                    objJS.add(c);
                }
            }
        }

        List list = new ArrayList();
        for (StoredObject c : objJS) {
            FileBean fb = new FileBean();
            fb.setName(c.getBareName());
            fb.setIsdirectory(c.isDirectory());
            if (!c.isDirectory()) {
                String temppath = "temp" + File.separator + c.getBareName();
                //StoredObject t = c.getAsObject();
                // downfile(upath, temppath,t);
                fb.setFilepath(temppath);
            }
            fb.setPath(c.getName());
            fb.setLength(c.getAsObject().getContentLength());
            fb.setLastmodified(DateUtil.gmtToHMS(c.getAsObject().getLastModified()));
```

```
            list.add(fb);
        }
        return list;
    }

    private Collection list(Container container) {
        return container.list();
    }
```

（2）在“src/com/xiandian/cloud/storage/service/SwiftStorageService.java”中增加如下代码：

```
/**
     * 取得路径下对应类型的所有文件
     *
     * @param email
     * @return
     */
    public List getCategoryStoredList(String username, String password,
            int type, String upath) {
        Account account = SwiftUtilTools.getAccount(username, password);
        SwiftStoreImpl Swiftdfs = new SwiftStoreImpl(account);
        List categoryStoredList = Swiftdfs
                .getCategoryStoredList(username, type);
        return categoryStoredList;
    }
```

**3. 实现控制层**

在“/src/com/xiandian/cloud/storage/Web/StorageController.java”类中增加如下代码：

```
/**
     * 取得特定类型的文件
     * @param request
     * @param response
     * @param type
     *          1 图片格式 2 文档格式 3 视频格式 4 音乐格式 5 其他
     * @return
     */
    @RequestMapping("/category")
    public ModelAndView category(HttpServletRequest request,
            HttpServletResponse response, int type) {
```

```
        ModelAndView view = new ModelAndView();
        User user = getSessionUser(request);
        String upath = request.getSession().getServletContext()
                .getRealPath("/");
        List list = storageService.getCategoryStoredList(user.getUsername(),
                user.getPassword(), type, upath);
        view.addObject("list", list);
        view.addObject("search", "true");
        view.addObject("type", type);
        view.setViewName("/main");
        return view;
    }
```

4. 实现前后台数据交互

根据控制层“StorageController”类展示的文件筛选“Category”方法，可以看到，前端实现数据交互只需向后端传输相应的文件类型 type，后台根据文件类型 type 向前端返回相应数据，然后显示在“Main.jsp”界面上。因此本案例为左侧导航栏“我的文档”“我的图片”“我的视频”“我的音乐”“其他”添加相应链接来实现数据交互，具体代码如下：

```
<div id="ahref">
 <a class="list-group-item active" id="r_active" href="home.action">
 <span class="glyphicon glyphicon-home left-icon"></span>
    全部文件</a>
 <a class="list-group-item " id="type" href="category.action?type=2">
 <span class="glyphicon glyphicon-file left-icon"> </span>
  我的文档</a>
 <a class="list-group-item " href="category.action?type=1">
 <span class="glyphicon glyphicon-picture left-icon"></span>
  我的图片</a>
 <a class="list-group-item " id="type" href="category.action?type=3" >
 <span class="glyphicon glyphicon-film left-icon"></span>
  我的视频</a>
 <a class="list-group-item " id="type" href="category.action?type=4" >
 <span class="glyphicon glyphicon-music left-icon"></span>
  我的音乐</a>
 <a class="list-group-item" id="type" href="category.action?type=5">
 <span class="glyphicon glyphicon-inbox left-icon"></span>
  其他</a>
```

```
<a class="list-group-item" id="type" href="shareFile.action">
<span class="glyphicon glyphicon-share left-icon"></span>
    我的分享</a>
<a class="list-group-item" href="garbage.action">
<span class="glyphicon glyphicon-trash left-icon"></span>
    垃圾箱</a>
</div>
```

#### 5. 功能验证测试

将项目部署到 Tomcat 中并发布，项目运行成功后，登录进入，单击侧边栏中的“我的图片”项，显示所有的图片文件，如图 6-5 所示。

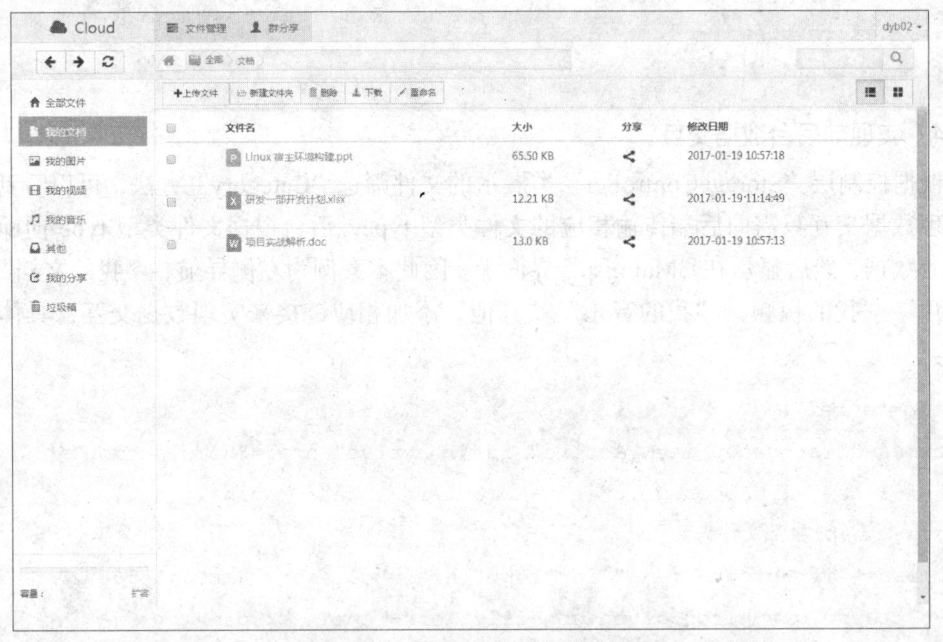


图 6-5　文件筛选分类功能运行效果图

本功能的测试场景见表 6-3。

表 6-3　文件筛选分类功能测试场景

| 编号 | 测试场景 | 输入参数 | 预期结果 |
|---|---|---|---|
| 1 | 登录成功后 | 单击侧边栏中“我的图片”项 | 显示图片列表视图 |

## 任务 6.4　开发文件缩略图显示功能

单击界面右上方的两个缩略图按钮，可以进行列表视图和网格视图的切换。

### 6.4.1　相关知识

本功能实现的技术原理如下。

（1）视图层：显示界面，如图 6-6 所示。

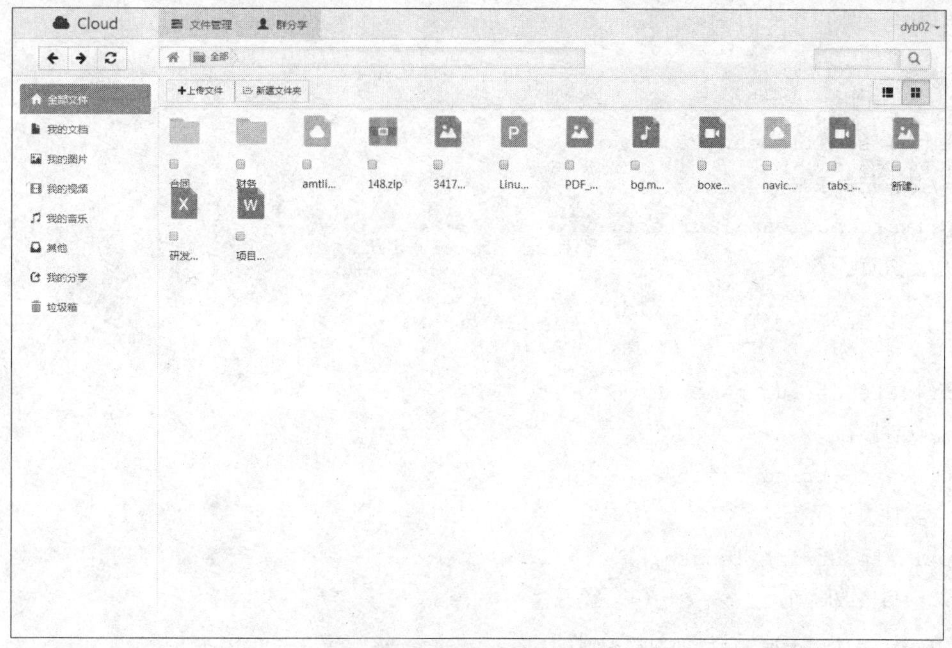


图 6-6　缩略图展示界面

（2）控制层：接收视图层的消息，向服务层发送文件缩略显示参数，并将服务层返回的结果发送到视图层。

（3）服务层：重新获取文件，将文件信息发送到控制层。

本功能的具体实现流程如图 6-7 所示。

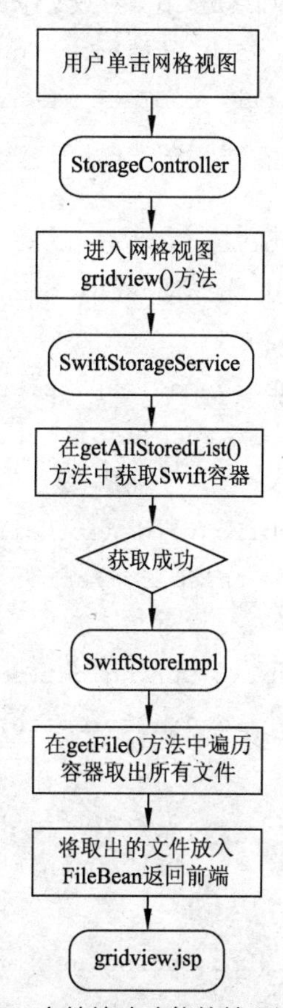


图 6-7　文件缩略功能的处理流程图

## 6.4.2　实现步骤

### 1．项目导入

在 Eclipse 中导入工程“project63”。

### 2．实现服务层

缩略图实现步骤与文件列表展示步骤类似。

（1）在 src/com/xiandian/cloud/storage/sh/bean 目录下新建一个名为 FileBean 的类，该类定义了文件的一些属性，包括文件名称、文件大小、文件路径、文件修改时间、是否为目录、是否有子文件夹等，这些信息将会被传给前台页面显示，具体代码如下：

```
import com.xiandian.cloud.storage.sh.util.UtilTools;
/**
 * 与前台交互数据的 bean
```

```
 * @author 云计算应用与开发项目组
 * @since  V1.4
 *
 */
public class FileBean {
    //是否是目录
    private boolean isdirectory;
    //是否有子文件夹
    private boolean haschild;
    //文件名称
    private String name;
    //文件路径
    private String path;
    //文件大小
    private String length;
    //最后修改时间
    private String lastmodified;
    //如果是文件，存放文件的临时路径
    private String filepath;
    public boolean isHaschild() {
        return haschild;
    }
    public void setHaschild(boolean haschild) {
        this.haschild = haschild;
    }
    public boolean isIsdirectory() {
        return isdirectory;
    }
    public void setIsdirectory(boolean isdirectory) {
        this.isdirectory = isdirectory;
    }
    public String getFilepath() {
        return filepath;
    }
    public void setFilepath(String filepath) {
        this.filepath = filepath;
    }
    public String getLength() {
        return length;
```

```
    }
    public void setLength(long length) {
        this.length = UtilTools.convertFileSize(length);
    }
    public String getLastmodified() {
        return lastmodified;
    }
    public void setLastmodified(String lastmodified) {
        this.lastmodified =lastmodified;
    }
    public String getName() {
        return name;
    }
    public void setName(String name) {
        this.name = name;
    }
    public String getPath() {
        return path;
    }
    public void setPath(String path) {
        this.path = path;
    }
}
```

（2）在"src/com/xiandian/cloud/storage/sh/SwiftStoreImpl.java" 类中增加如下代码：

```
/**
     * 获取文件类表
     *
     * @param rpath 当前目录
     * @return List
     */
    public List getFile(String rpath) {
        Map mappath = getSplitPath(rpath);
        String path = mappath.get("path").toString();
        Container container = account.getContainer(mappath.get("rootPath")
                .toString());
        Collection<DirectoryOrObject> objects;
        if (path != null && path.length() > 1) {
            Directory directory = new Directory(path, '/');
```

```
            objects = listDirectory(container, directory);
        } else {
            objects = listDirectory(container);
        }
        List list = new ArrayList();
        for (DirectoryOrObject c : objects) {
            FileBean fb = new FileBean();
            fb.setName(c.getBareName());
            fb.setIsdirectory(c.isDirectory());
            if (c.isDirectory()) {
                fb.setPath(c.getAsDirectory().getName());
                list.add(fb);
            } else {
                StoredObject t = c.getAsObject();
                String tp = t.getPath();
                String ttp = SwiftUtilTools.replaceStr(tp);
                if (!ttp.endsWith("/")) {
                    fb.setPath(t.getName());
                    fb.setLength(t.getContentLength());
                    fb.setLastmodified(t.getLastModified());
                    list.add(fb);
                }
            }
        }
        return list;
    }
    private Map<String, String> getSplitPath(String path) {
        String[] strPath = path.split("/");
        String rootPath = "";
        path = "";
        for (int i = 0; i < strPath.length; i++) {
            if (i == 0) {
                rootPath = strPath[0];
            } else {
                if (i == strPath.length - 1) {
                    if (strPath[i].contains(".")) {
                        path += strPath[i];
                    } else {
                        path += strPath[i] + "/";
```

```
                }
            } else {
                path += strPath[i] + "/";
            }
        }
    }
    Map map = new HashMap();
    map.put("rootPath", rootPath);
    map.put("path", path);
    return map;
}
private Collection listDirectory(Container container, Directory directory)
{
    return container.listDirectory(directory);
}
private Collection listDirectory(Container container) {
    return container.listDirectory();
}
```

（3）在 com/xiandian/cloud/storage/util 目录下新建一个名为 ComparatorFile 的类，该类对文件和文件夹进行排序，由界面调用，在文件列表中，先显示文件夹再显示文件。具体代码如下：

```
package com.xiandian.cloud.storage.util;
import java.util.Comparator;
import com.xiandian.cloud.storage.sh.bean.FileBean;
/**
 * 文件夹与文件排序的实现类
 *
 */
public class ComparatorFile implements Comparator<FileBean> {
    public ComparatorFile() {
    }
    public int compare(FileBean f1, FileBean f2) {
        boolean f1dir = f1.isIsdirectory();
        boolean f2dir = f2.isIsdirectory();
        if (f1dir && f2dir) {
            return 0;
        } else if (!f1dir && f2dir) {
            return 1;
```

```
        } else if (f1dir && !f2dir) {
            return -1;
        } else if (!f1dir && !f2dir) {
            return 0;
        }
        return 0;
    }
}
```

（4）在“src/com/xiandian/cloud/storage/service/ SwiftStorageService.java”类中增加如下代码：

```
/**
     * 取得路径下的所有文件
     */
    public List getAllStoredList(String username, String password, String path)
{
        Account account = SwiftUtilTools.getAccount(username, password);
        SwiftStoreImpl Swiftdfs = new SwiftStoreImpl(account);
        String up = null;
        up = username + "/" + path;
        List list = Swiftdfs.getFile(up);
        ComparatorFile comparator = new ComparatorFile();
        if (!list.isEmpty()) {
            synchronized (list) {
                Collections.sort(list, comparator);
            }

        }
        return list;
}
```

### 3. 实现控制层

在“/src/com/xiandian/cloud/storage/Web/StorageController.java”中增加如下代码：

```
/**
     * 取得主页的缩略图
     *
     * @param request
     * @param response
     * @param path
```

```
     * @return
     */
    @RequestMapping("/gridview")
    public ModelAndView gridview(HttpServletRequest request,
            HttpServletResponse response, String path) {
        ModelAndView view = new ModelAndView();
        User user = getSessionUser(request);
        path = UtilTools.converStr(path);
        List list = storageService.getAllStoredList(user.getUsername(),
                user.getPassword(), path);
        view.addObject("path", path);
        view.addObject("list", list);
        view.setViewName("/gridview");
        return view;
    }
```

4. 实现前后台数据交互

在控制层“StorageController”类展示缩略图方法 GridView 中，可以看到后台列表数据是跟随页面一起返回的并显示在“gridview.jsp”中，因此只需通过“jstl <c:foreach>”方法来循环取值，并将内容放置在“id="gridview"”的代码块中，具体代码如下：

```
<c:forEach var="fb" items="${list}" varStatus="status">
 <div class=" col-lg-1 col-md-1 col-sm-2 col-xs-12 gridview-imgbox">
 <c:choose>
 <c:when test="${fb.isdirectory == true}">
 <a href="gridview.action?path=${fb.path}">
 <img src="assets/images/Folder_lg.png" alt="No image" class="gridview-img">
 </a>
 </c:when>
 <c:when test="${fn:endsWith(fb.name,'.txt') || fn:endsWith(fb.name,'.TXT')}">
 <img src="assets/images/Text_lg.png" class="gridview-img"></c:when>
<c:when
test="${fn:endsWith(fb.name,'.docx')||fn:endsWith(fb.name,'.DOCX')}">
 <img src="assets/images/Word_lg.png" class="gridview-img"></c:when>
 <c:when test="${fn:endsWith(fb.name,'.doc')||fn:endsWith(fb.name,'.DOC')}">
 <img src="assets/images/Word_lg.png" class="gridview-img"></c:when>
 <c:when
test="${fn:endsWith(fb.name,'.xls')||fn:endsWith(fb.name,'.xlsx')}">
 <img src="assets/images/Excel_lg.png" class="gridview-img"></c:when>
 <c:when
```

```
test="${fn:endsWith(fb.name,'.XLSX')||fn:endsWith(fb.name,'.XLS')}">
 <img src="assets/images/Excel_lg.png" class="gridview-img"></c:when>
 <c:when test="${fn:endsWith(fb.name,'.PPT')||fn:endsWith(fb.name,'.PPTX')}">
 <img src="assets/images/PPT_lg.png" class="gridview-img"></c:when>
 <c:when test="${fn:endsWith(fb.name,'.pptx')||fn:endsWith(fb.name,'.ppt')}">
 <img src="assets/images/PPT_lg.png" class="gridview-img"></c:when>
 <c:when test="${fn:endsWith(fb.name,'.pdf') || fn:endsWith(fb.name,'.PDF')}">
 <img src="assets/images/PDF_lg.png" class="gridview-img"></c:when>
 <c:when test="${fn:endsWith(fb.name,'.mp3')||fn:endsWith(fb.name,'.MP3')}">
 <img src="assets/images/Music_lg.png" class="gridview-img"></c:when>
 <c:when test="${fn:endsWith(fb.name,'.cda')||fn:endsWith(fb.name,'.mid')}">
 <img src="assets/images/Music_lg.png" class="gridview-img"></c:when>
 <c:when test="${fn:endsWith(fb.name,'.wav')||fn:endsWith(fb.name,'.WAV')}">
 <img src="assets/images/Music_lg.png" class="gridview-img"></c:when>
 <c:when test="${fn:endsWith(fb.name,'.mp4')||fn:endsWith(fb.name,'.MP4')}">
 <img src="assets/images/Video_lg.png" class="gridview-img"></c:when>
 <c:when test="${fn:endsWith(fb.name,'.wmv')||fn:endsWith(fb.name,'.WMV')}">
 <img src="assets/images/Video_lg.png" class="gridview-img"></c:when>
 <c:when test="${fn:endsWith(fb.name,'.rmvb')||fn:endsWith(fb.name,'.RMVB')}">
 <img src="assets/images/Video_lg.png" class="gridview-img"></c:when>
 <c:when test="${fn:endsWith(fb.name,'.swf')||fn:endsWith(fb.name,'.SWF')}">
 <img src="assets/images/Video_lg.png" class="gridview-img"></c:when>
 <c:when test="${fn:endsWith(fb.name,'.flv')||fn:endsWith(fb.name,'.FLV')}">
 <img src="assets/images/Video_lg.png" class="gridview-img"></c:when>
 <c:when test="${fn:endsWith(fb.name,'.avi')||fn:endsWith(fb.name,'.AVI')}">
 <img src="assets/images/Video_lg.png" class="gridview-img"></c:when>
 <c:when test="${fn:endsWith(fb.name,'.rar')||fn:endsWith(fb.name,'.RAR')}">
 <img src="assets/images/ZIP_lg.png" class="gridview-img"></c:when>
 <c:when test="${fn:endsWith(fb.name,'.zip')||fn:endsWith(fb.name,'.ZIP')}">
 <img src="assets/images/ZIP_lg.png" class="gridview-img"></c:when>
 <c:when test="${fn:endsWith(fb.name,'.jpg') || fn:endsWith(fb.name,'.JPG')}">
 <img src="assets/images/Picture.png" class="gridview-img"></c:when>
 <c:when test="${fn:endsWith(fb.name,'.png')||fn:endsWith(fb.name,'.PNG')}">
 <img src="assets/images/Picture.png" class="gridview-img"></c:when>
 <c:when test="${fn:endsWith(fb.name,'.gif')||fn:endsWith(fb.name,'.GIF')}">
 <img src="assets/images/Picture.png" class="gridview-img"></c:when>
 <c:when test="${fn:endsWith(fb.name,'.jpeg')||fn:endsWith(fb.name,'.JPEG')}">
 <img src="assets/images/Picture.png" class="gridview-img"></c:when>
 <c:when test="${fn:endsWith(fb.name,'.ico')||fn:endsWith(fb.name,'.ICO')}">
```

```
<img src="assets/images/Picture.png" class="gridview-img"></c:when>
<c:otherwise>
<img src="assets/images/Other_lg.png" class="gridview-img"></c:otherwise>
</c:choose>
<div class="text">
<div class="content">
<div style="display: none">${fb.filepath}</div>
<div style="display: none">${fb.path}</div>
<p class="p_len">
<input type="checkbox" class="checkbox gridview-checkbox"
  name="check"onclick="show()"id="input${status.count}"value="${fb.name}"
    style="float:left"/>
<span class="p_lean1" ></span></p></div></div></div></c:forEach>
```

5. 功能验证测试

将项目部署到 Tomcat 中并发布，项目运行成功后，登录进入，单击菜单栏的“切换视图”按钮，以网格视图显示所有文件，如图 6-8 所示。

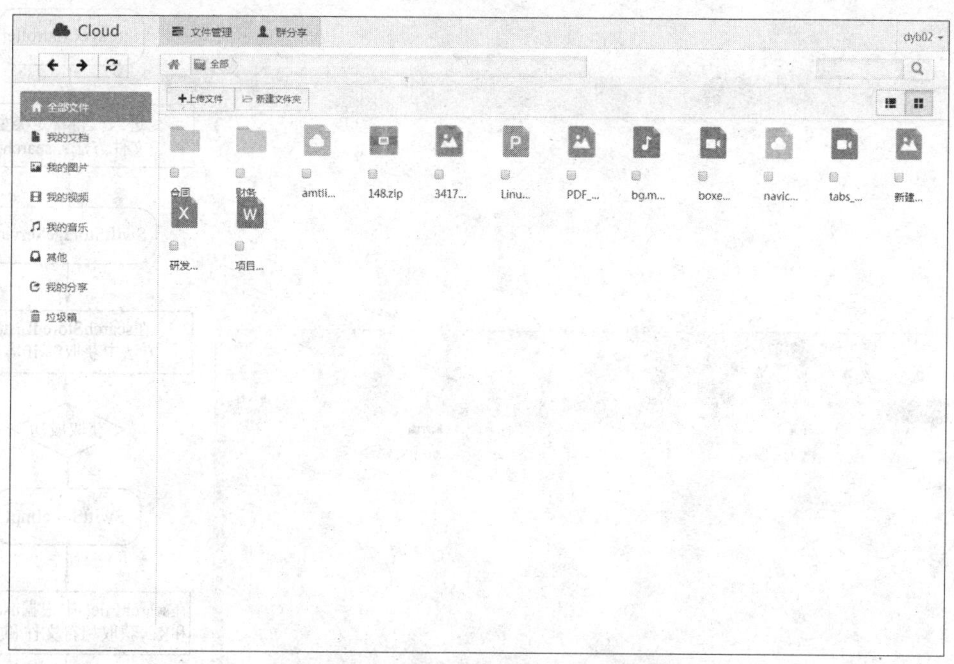


图 6-8　文件缩略图功能运行效果图

本功能的测试场景见表 6-4。

表 6-4　文件缩略图功能测试场景

| 编号 | 测试场景 | 输入参数 | 预期结果 |
|---|---|---|---|
| 1 | 登录成功后 | 单击网格模式显示按钮 | 以网格方式显示所有文件 |
| 2 | 登录成功后 | 单击列表模式显示按钮 | 以列表方式显示所有文件 |

## 任务 6.5 开发文件搜索功能

在右上角“搜索输入框”中输入文件名称，单击“搜索”按钮，并展示符合搜索条件的文件列表。

### 6.5.1 相关知识

本功能实现的技术原理如下。

（1）视图层：显示界面如图 6-9 所示。

（2）控制层：接收视图层的消息，向服务层发送文件搜索参数，并将服务层返回的结果发送到视图层。

（3）服务层：根据关键词搜索文件，将符合搜索条件的文件封装到集合中，返回给控制层。

本功能的具体实现流程如图 6-10 所示。

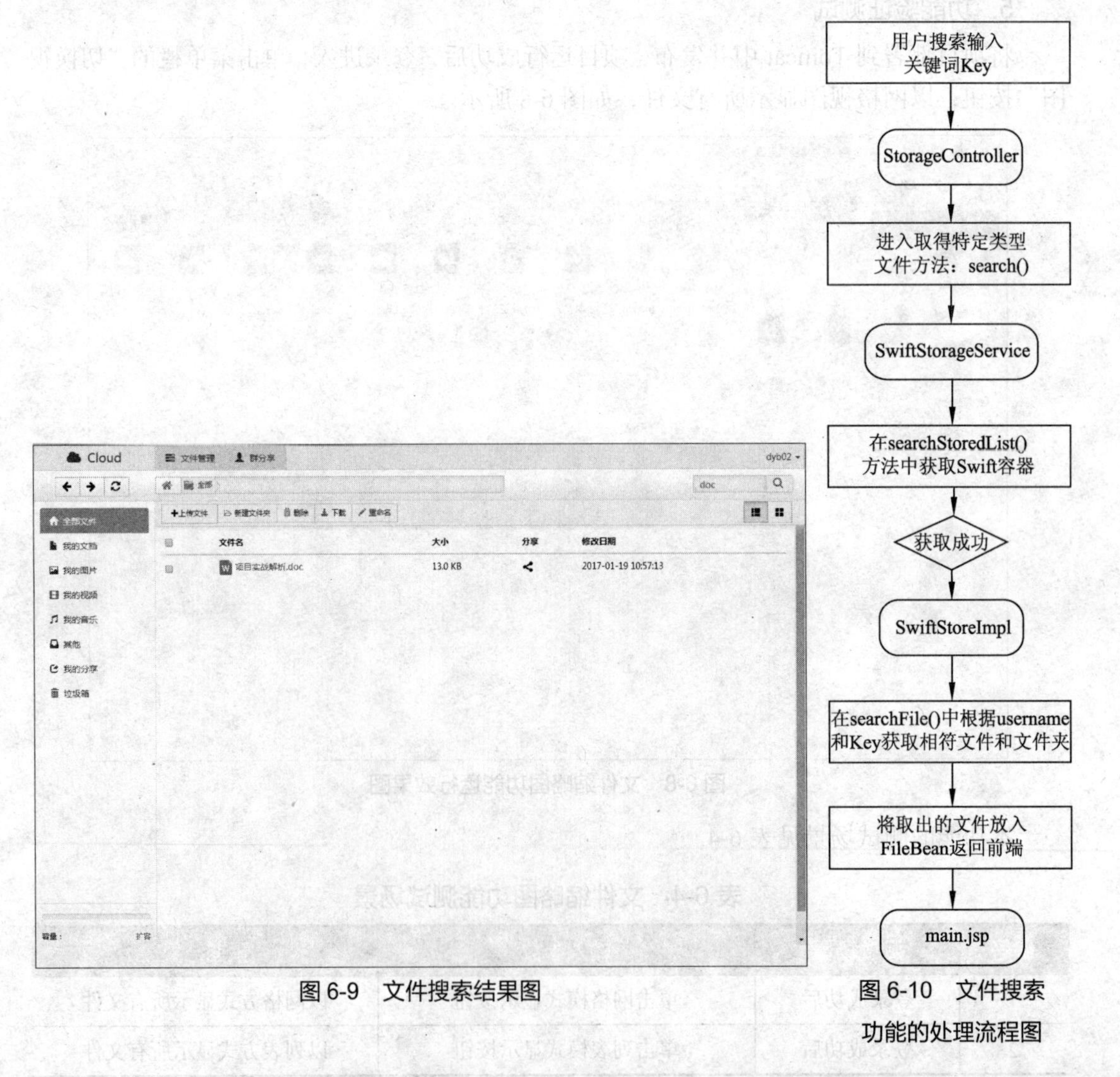


图 6-9 文件搜索结果图

图 6-10 文件搜索功能的处理流程图

### 6.5.2 实现步骤

**1. 项目导入**

在“Eclipse”中导入工程“project64”。

**2. 实现服务层**

在“src/com/xiandian/cloud/storage/service/SwiftStorageService.java ”中增加如下代码：

```
/**
 * 搜索根路径下的所有与 key 匹配的文件及文件夹
 *
 * @param username
 * @param upath
 * @param key
 * @return
 */
public List searchStoredList(String username, String password,
        String upath, String key) {
    List list = new ArrayList();
    Account account = SwiftUtilTools.getAccount(username, password);
    SwiftStoreImpl Swiftdfs = new SwiftStoreImpl(account);
    list = Swiftdfs.searchFile(username, key);
    ComparatorFile comparator = new ComparatorFile();
    if (!list.isEmpty()) {
        synchronized (list) {
            Collections.sort(list, comparator);
        }
    }
    return list;
}
```

**3. 实现控制层**

（1）在“src/com/xiandian/cloud/storage/sh/SwiftStoreImpl.java ”中增加如下代码：

```
/**
 * 描述：搜索文件
 *
 * @param rootPath 根路径
 * @param key 搜索条件
 * @return List
 */
public List searchFile(String rootPath, String key) {
        Container container = account.getContainer(rootPath);
```

```
        Collection<StoredObject> ll = list(container);
        List list = new ArrayList();
        for (StoredObject c : ll) {
            String path = c.getName();
            String name = c.getBareName();
            if (name != null
                    && name.toLowerCase().matches(
                            ".{0,}" + key.toLowerCase() + ".{0,}")) {
                FileBean fb = new FileBean();
                fb.setName(name);
                boolean isdir = path.endsWith("/");
                fb.setIsdirectory(isdir);
                if (isdir) {
                    fb.setPath(c.getName());
                } else {
                    String temppath = "temp" + File.separator + c.getBareName();
                    fb.setFilepath(temppath);
                    fb.setPath(path);
                    fb.setLength(c.getContentLength());
                    fb.setLastmodified(c.getLastModified());
                }
                list.add(fb);
            }
        }
        return list;
    }
```

（2）在“/src/com/xiandian/cloud/storage/Web/StorageController.java”中增加如下代码：

```
/**
     * 搜索文件
     *
     * @param request
     * @param response
     * @param path
     * @return
     */
    @RequestMapping("/search")
    public ModelAndView search(HttpServletRequest request,
            HttpServletResponse response, String key) {
        ModelAndView view = new ModelAndView();
        User user = getSessionUser(request);
        String upath = request.getSession().getServletContext()
                .getRealPath("/");
```

```
        key = UtilTools.converStr(key);
        List list = storageService.searchStoredList(user.getUsername(),
                user.getPassword(), upath, key);
        view.addObject("list", list);
        view.addObject("search", "true");
        view.setViewName("/main");
        return view;
    }
```

### 4. 实现前后台数据交互

在控制层“StorageController”类的查询方法“search”中，可以看到该方法需要向后台传输一个“Key”才能返回对应列表，因此前端需要获取用户输入的值并将该值传给后台即可获得数据返回。找到“id="searchfiles"”的“<a>”标签，同时为其添加“onclick”事件“searchfile()”，并在 js 中填写相应代码实现交互：

```
function searchfile() {
    var key = $("#navbarInput-01").val();
    $("#searchfiles").attr("href", "search.action?key=" + key);
}
```

### 5. 功能验证测试

将项目部署到 Tomcat 中并发布，项目运行成功后，登录进入，在搜索输入框中输入搜索文件名称并回车，显示符合搜索条件的文件列表，运行效果如图 6-11 所示。

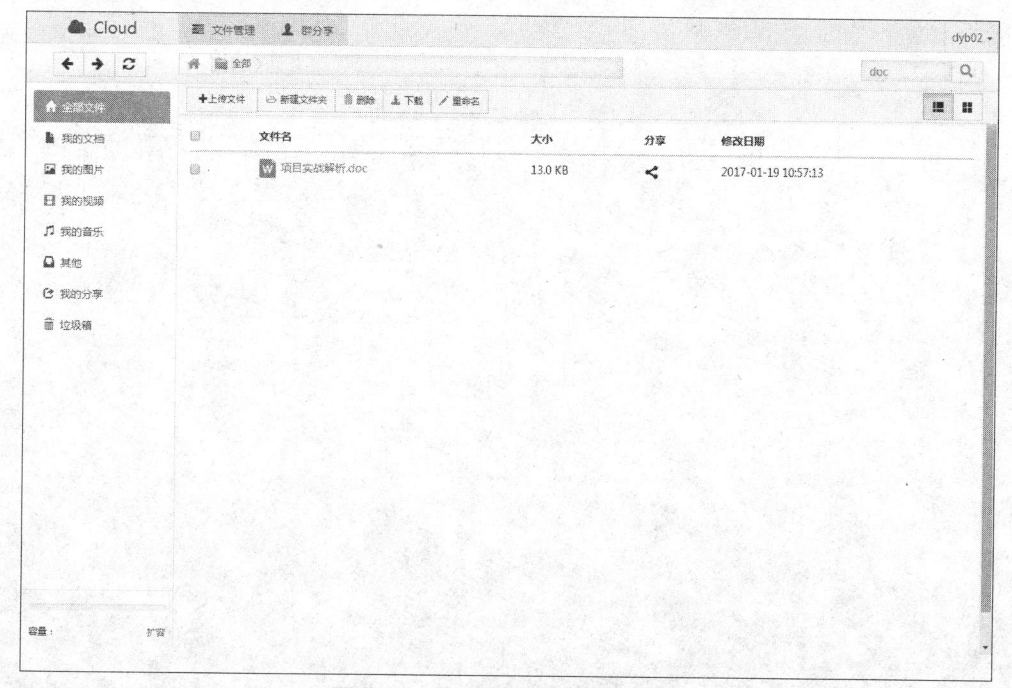


图 6-11　文件搜索功能运行效果图

本功能的测试场景见表 6-5。

表 6-5　文件搜索功能测试场景

| 编号 | 测试场景 | 输入参数 | 预期结果 |
| --- | --- | --- | --- |
| 1 | 登录成功后 | jpg | 显示所有图片文件 |
| 2 | 登录成功后 | /\（特殊字符） | 显示“找不到此文件” |

# 项目 7 开发文件操作模块

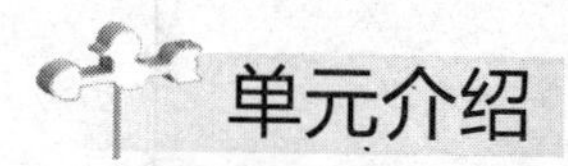

## 单元介绍

本单元读者需要掌握网盘文件操作模块的功能实现，包括前后台代码的流程，并对其功能进行验证测试。

## 学习任务

本单元主要完成以下学习目标：

- 掌握网盘文件和文件夹的创建、重命名、复制和粘贴等功能。

根据这些学习目标，本项目将分解为以下 4 个任务，见表 7-1。

表 7-1 任务分解表

| 任务序号 | 任务内容 | 验证方式 |
| --- | --- | --- |
| 任务 7.1 | 开发文件夹的创建功能 | 可以创建新的文件夹 |
| 任务 7.2 | 开发文件夹和文件的重命名功能 | 可以对指定文件和文件夹重命名 |
| 任务 7.3 | 开发文件复制和粘贴功能 | 可以复制文件到指定目录 |
| 任务 7.4 | 开发文件移动功能 | 可以将文件移动到指定目录 |

## 任务 7.1 开发文件夹的创建功能

用户单击“新建文件夹”按钮，弹出对话框，用户输入文件夹名称，单击“确定”按钮后，在当前目录下增加对应的文件夹。

### 7.1.1 相关知识

本功能实现的技术原理如下。

（1）视图层：显示界面如图 7-1 所示。

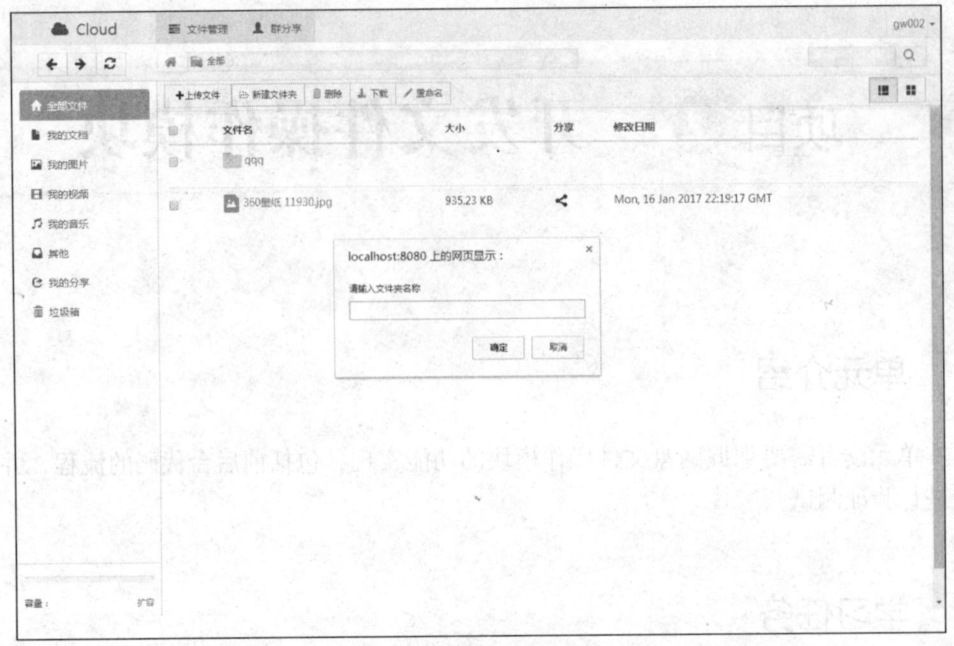


图 7-1　创建文件夹操作

（2）控制层：接收视图层的消息，向服务层发送文件夹创建参数，并将服务层返回的结果发送到视图层。

（3）服务层：调用 Swift 服务创建目录，并将创建结果返回给控制层。

本功能的具体实现流程如图 7-2 所示。

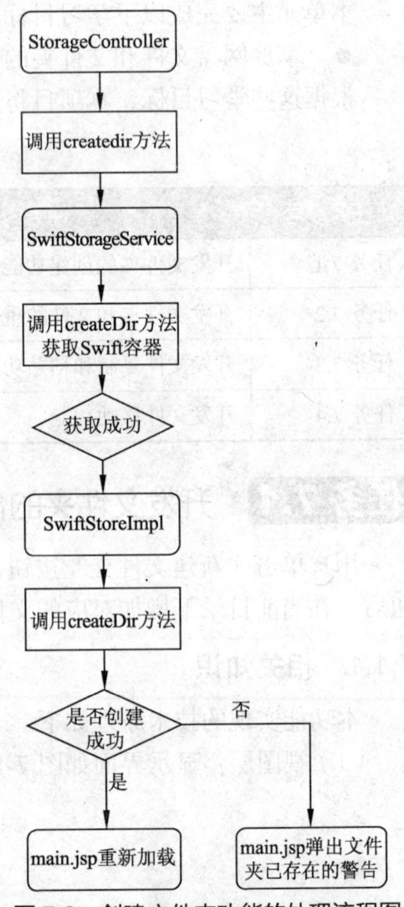


图 7-2　创建文件夹功能的处理流程图

### 7.1.2　实现步骤

#### 1. 项目导入

在 Eclipse 中导入工程“project65”。

#### 2. 实现服务层

（1）在“src/com/xiandian/cloud/storage/service/SwiftStorageService.java”中增加如下代码：

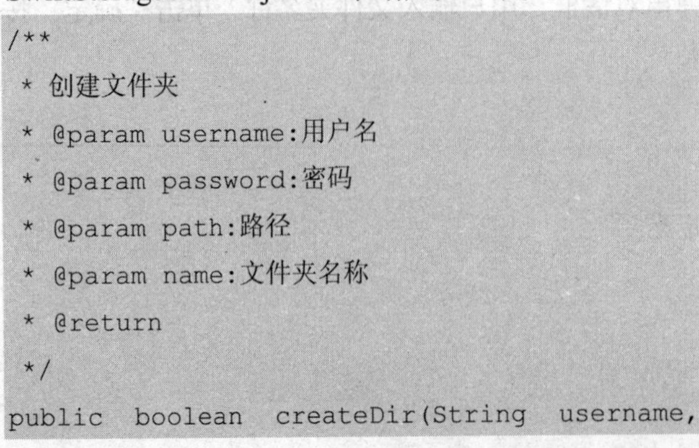

```
/**
 * 创建文件夹
 * @param username:用户名
 * @param password:密码
 * @param path:路径
 * @param name:文件夹名称
 * @return
 */
public  boolean  createDir(String  username,
```

```
String password, String path,String name) {
    Account account = SwiftUtilTools.getAccount(username, password);
    SwiftStoreImpl Swiftdfs = new SwiftStoreImpl(account);
    Swiftdfs.createDir(username + "/" + path, name);
    return true;
}
```

（2）在“src/com/xiandian/cloud/storage/sh/SwiftStoreImpl.java”中增加如下代码：

```
/**
 * 创建文件夹
 * @param rpath:文件夹的主目录
 * @param name:文件夹名称
 * @return
 */
public boolean createDir(String rpath, String name) {
    Map mappath = getSplitPath(rpath);
    String path = mappath.get("path").toString();
    Container container = account.getContainer(mappath.get("rootPath").toString());
    if (path == null) {
        path = "";
    }
    StoredObject object = getObject(container, path + name + "/");
    boolean isexist = object.exists();
    if (isexist) {
        return false;
    }
    uploadObject(object);
    return true;
}
private StoredObject getObject(Container container, String path) {
    return container.getObject(path);
}
private void uploadObject(StoredObject newObject) {
    newObject.uploadObject(new byte[] {});
}
```

### 3. 实现控制层

在“/src/com/xiandian/cloud/storage/Web/StorageController.java”中增加如下代码：

```
/**
    * 创建文件夹
```

```
     *
     * @param request:请求
     * @param path:路径
     * @param name:文件夹名称
     */
    @RequestMapping("/createdir")
    @ResponseBody
    public Object createDir(HttpServletRequest request,String path, String name) {
        User user = getSessionUser(request);
        boolean flag = false;
        flag = storageService.createDir(user.getUsername(), user.getPassword(),
path,name);
        return new MessageBean(flag, flag ? Constants.SUCCESS_3: Constants.ERROR_4);
}
```

4. 实现前后台数据交互

为“新建文件夹”按钮绑定单击事件“newdir()”，用户通过单击按钮触发添加文件夹操作。

```
//校验文件夹名
function checkfolder(name){
    var reg = new RegExp('^[^\\\\/:*?"<>|%]+$');
    var flag = reg.test(name);
    if(!flag){
        alert("文件名不能包含以下字符：<,>,|,*,?,\\,/,%");
    };
    return flag;
}
//新建文件夹
function newdir() {
    var name = prompt("请输入文件夹名称", '');
    if(name==""){
        alert("文件名不能为空！");
        return false;
    }
    if(name==null){
        return false;
    }
    if(!checkfolder(name)){
        return false;
    }
        var dirname = name;
```

```
        var path = "${path}";
        var data = {
            path : path,
            name : dirname
        };
        $.ajax({
            url : "createdir.action",
            type : "post",
            data : data,
            success : function(s) {
                if (s.success) {
                    location.reload();
                } else {
                    alert(s.msg);
                };
            }
        });
}
```

5. 功能验证测试

将项目部署到 Tomcat 中并发布，项目运行成功后，登录进入，单击菜单栏的“新建文件夹”按钮，效果如图 7-3 所示。

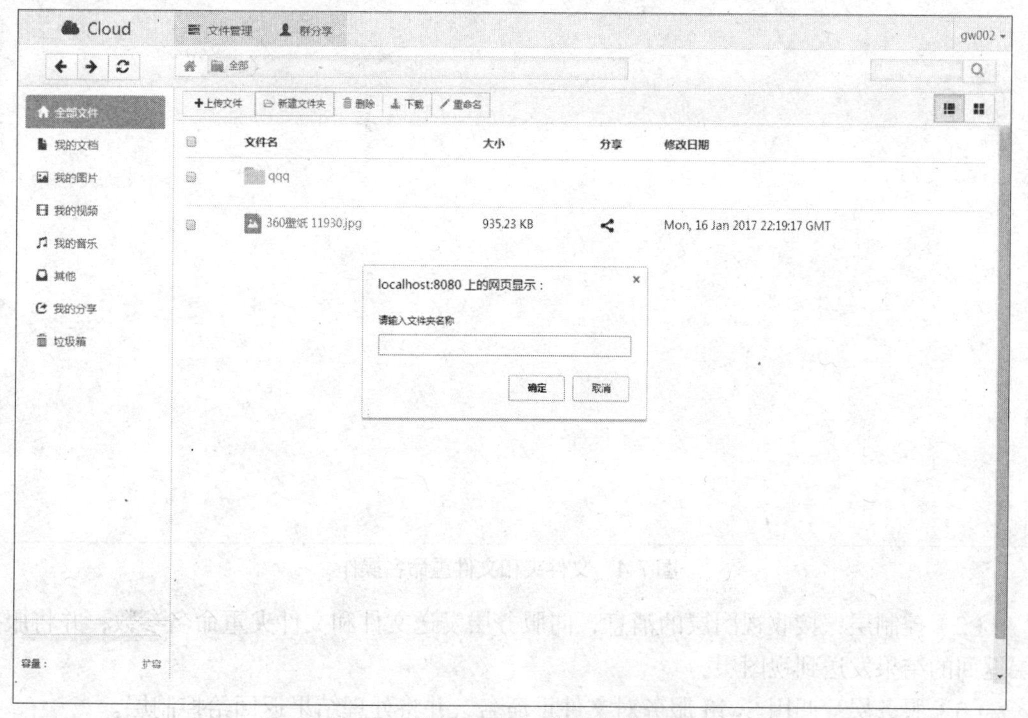


图 7-3　文件夹创建功能运行效果图

本功能的测试场景见表 7-2。

表 7-2　文件夹创建功能测试场景

| 编号 | 测试场景 | 输入参数 | 预期结果 |
| --- | --- | --- | --- |
| 1 | 文件名不重复，不含特殊字符 | dirName：wwww | 创建成功 |
| 2 | 文件名重复，不含特殊字符 | dirName：wwww | 提示目录名称重复 |
| 3 | 文件名不重复，含特殊字符 | dirName：text/ww | 提示目录名称不合法 |

## 任务 7.2　开发文件夹和文件的重命名功能

用户选中某一文件或文件夹后，单击“重命名”按钮，弹出对话框，在对话框中输入新的文件名称后，单击“确定”按钮，修改对应文件或文件夹的名称。

### 7.2.1　相关知识

本功能实现的技术原理如下。

（1）视图层：显示界面如图 7-4 所示。

图 7-4　文件夹和文件重命名操作

（2）控制层：接收视图层的消息，向服务层发送文件和文件夹重命名参数，并将服务层返回的结果发送到视图层。

（3）服务层：调用 Swift 服务对文件重命名，并将处理结果返回给控制层。

本功能的具体实现流程如图 7-5 所示。

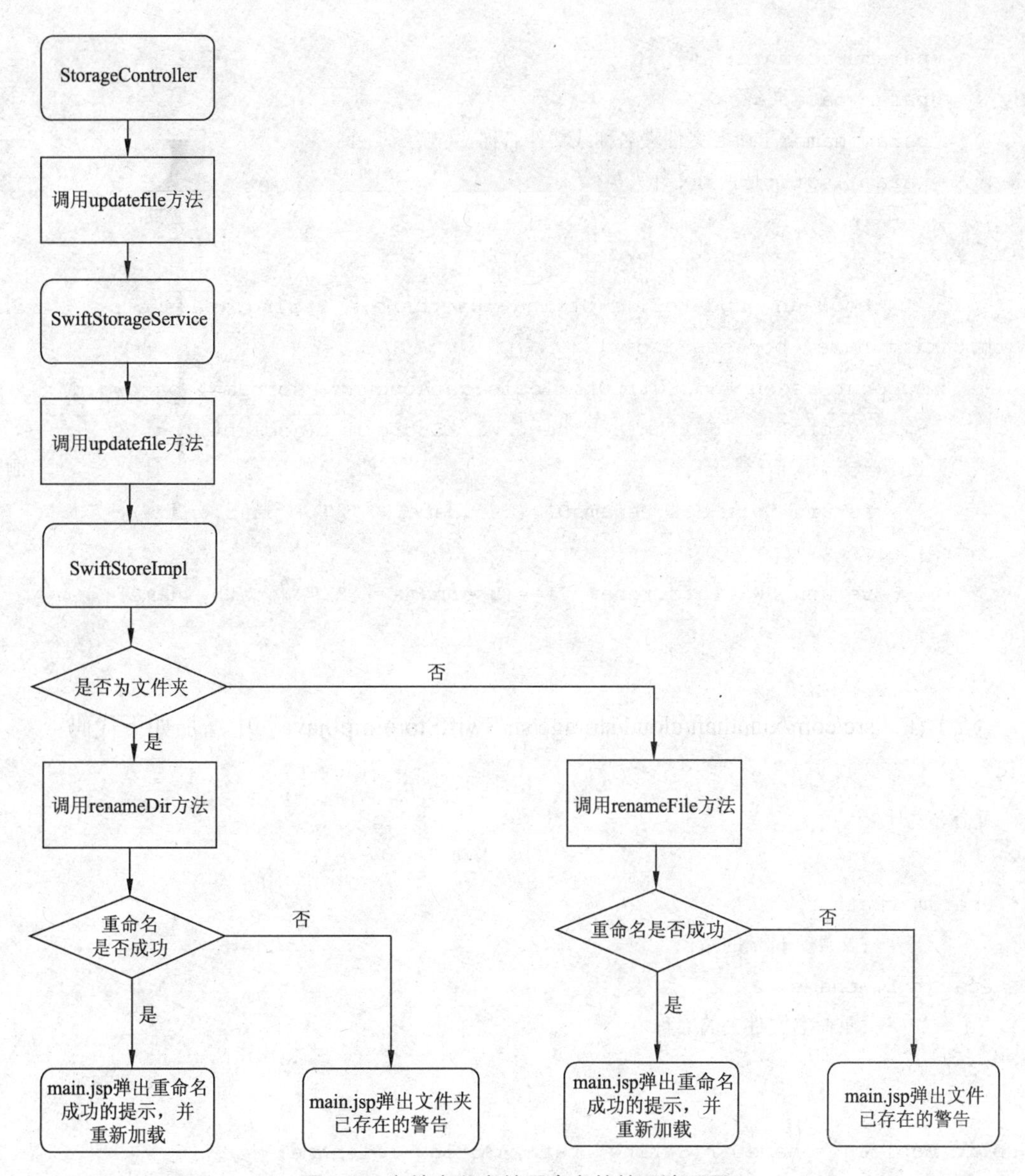


图 7-5　文件夹和文件重命名的处理流程图

### 7.2.2　实现步骤

#### 1. 项目导入

在 Eclipse 中导入工程“project71”。

#### 2. 实现服务层

（1）在“src/com/xiandian/cloud/storage/service/SwiftStorageService.java”中增加如下代码：

```
/**
     * 重命名文件
     * @param username:用户名
```

```
     * @param password:密码
     * @param path:路径
     * @param name:重命名文件夹名称或文件名称
     * @param isdir:是否为文件夹
     * @return
     */
    public boolean updateFile(String username, String password, String path,String name, boolean isdir) {
        Account account = SwiftUtilTools.getAccount(username, password);
        SwiftStoreImpl Swiftdfs = new SwiftStoreImpl(account);
        if (isdir) {
            return Swiftdfs.renameDir(username + "/" + path, name);
        } else {
            return Swiftdfs.renameFile(username + "/" + path, name);
        }
    }
```

（2）在“src/com/xiandian/cloud/storage/sh/SwiftStoreImpl.java”中增加如下代码：

```
/**
 * 重命名文件夹
 *
 * @param rpath
 *            :文件夹目录的路径
 * @param filename
 *            :重命名文件夹的名称
 * @return
 */
public boolean renameDir(String rpath, String filename) {
    Map mappath = getSplitPath(rpath);
    String path = mappath.get("path").toString();
    String rootpath = mappath.get("rootPath").toString();
    if (!path.contains("/")) {
        path = "/" + mappath.get("path").toString();
    }
    int len = rootpath.length();
    String temppath = SwiftUtilTools.replaceStr(path);
    String endstr = temppath.substring(0, temppath.length() - 1);
    int index = endstr.lastIndexOf("/");
    String newpath = null;
```

```
    if (index != -1) {
        newpath = temppath.substring(0, index + 1) + filename + "/";
    } else {
        newpath = filename + "/";
    }
    boolean flag = isexist(rootpath, newpath);
    if (flag) {
        rename(len, rootpath, path, path, newpath);
        renameDirFile (rootpath, path, newpath);
        return flag;
    }
    return flag;
}

private void rename(int len, String rootpath, String epath, String path,
        String newpath) {
    Container container = account.getContainer(rootpath);
    Directory directory = new Directory(epath, '/');
    Collection<DirectoryOrObject> objects = listDirectory(container,
            directory);
    for (DirectoryOrObject c : objects) {
        if (c.isDirectory()) {
            String cpath = c.getAsDirectory().getName();
            rename(len, rootpath, cpath, path, newpath);
            renamecDirFile (rootpath, cpath, path, newpath);
        } else {
            String temppath = c.getName();
            renamecfilefile(len, rootpath, temppath, path, newpath);
        }
    }
}

private void renameDirFile(String rootpath, String path, String newpath) {
    Container container = account.getContainer(rootpath);
    StoredObject newObject = getObject(container, newpath);
    uploadObject(newObject);
    StoredObject object = getObject(container, path);
    delete(object);
}
```

```
private void renamecDirFile(String rootpath, String cpath, String path,
        String newpath) {
    Container container = account.getContainer(rootpath);
    String temppath = SwiftUtilTools.replaceStr(cpath);
    String tpath = temppath.substring(path.length());
    StoredObject newObject = getObject(container, newpath + tpath);
    uploadObject(newObject);
    StoredObject object = getObject(container, temppath);
    delete(object);
}

/**
 * 文件重命名
 *
 * @param rpath
 *            : 重命名文件的路径
 * @param name
 *            : 重命名文件名称
 * @return
 */
public boolean renameFile(String rpath, String name) {
    Map mappath = getSplitPath(rpath);
    String path = mappath.get("path").toString();
    Container container = account.getContainer(mappath.get("rootPath")
            .toString());
    if (!path.contains("/")) {
        path = mappath.get("path").toString();
    }
    String temppath = SwiftUtilTools.replaceStr(path);
    StoredObject object = getObject(container, temppath);
    String tmp = temppath.substring(0, temppath.lastIndexOf("/") + 1);
    StoredObject newObject = getObject(container, tmp + name);
    boolean isexist = newObject.exists();
    if (isexist) {
        return false;
    }
    copyObject(object, container, newObject);
    delete(object);
```

```
    return true;
}

private void renamecfilefile(int len, String rootpath, String cpath,
        String path, String newpath) {
    Container container = account.getContainer(rootpath);
    String temppath = SwiftUtilTools.replaceStr(cpath);
    if (temppath.endsWith("/")) {
        return;
    }
    String tpath = temppath.substring(path.length());
    StoredObject newObject = getObject(container, newpath + tpath);
    StoredObject object = getObject(container, temppath);
    copyObject(object, container, newObject);
    delete(object);
}

private boolean isExist(String rootpath, String newpath) {
    Container container = account.getContainer(rootpath);
    StoredObject newObject = getObject(container, newpath);
    boolean isexist = newObject.exists();
    if (isexist) {
        return false;
    }
    return true;
}
private void copyObject(StoredObject object, Container container,
        StoredObject newObject) {
    object.copyObject(container, newObject);
}

private void delete(StoredObject object) {
    object.delete();
}
```

3. 实现控制层

在“/src/com/xiandian/cloud/storage/Web/StorageController.java”中增加如下代码：

```
/**
   * 重命名文件
```

```
     * @param request:请求
     * @param path:路径
     * @param name:文件夹名称或文件名称
     * @param isDir:是否为文件夹
     * @return
     */
    @RequestMapping("/updatefile")
    @ResponseBody
    public Object updateFile(HttpServletRequest request, String path, String
name,boolean isDir) {
        User user = getSessionUser(request);
       boolean flag = storageService.updateFile(user.getUsername(),user.getPassword(),
path, name, isDir);
        return new MessageBean(flag, flag ? Constants.SUCCESS_7: (isDir ?
Constants.ERROR_4 : Constants.ERROR_5));
    }
```

4. 实现前后台数据交互

当用户选中某一文件后，单击“重命名”按钮，弹出编辑框，用户输入新的文件名，然后单击“保存”按钮，完成重命名。因此前端的实现也分为两个步骤。

（1）获取选中文件，显示输入框。

① 用户选择某一文件，触发单击事件“show()”。

```
//显示重命名、下载、删除、移动、复制、分享选项
 function show() {
 var objTable = document.getElementById("tab");
 for (var i = 0; i < objTable.rows.length; i++) {
 var checkbox = objTable.rows[i].childNodes[1].childNodes[0].childNodes[5];
 if (checkbox.checked == true) {
document.getElementById("rename").style.display = "block";
break;
 } else {
document.getElementById("rename").style.display = "none";
    }
 }
}
```

② 为“重命名”按钮绑定单击事件“rename()”。

```
<button class="btn btn-default" type="button" id="rename" onclick="rename()">
 <i class="font-icon glyphicon glyphicon-pencil" style="color: #888"></i>重命名
</button>
```

③ 获取选中的文件，显示输入框。

```
//显示重命名框
 function rename() {
 var num = 0;
 var objTable = document.getElementById("tab");
 for (var y = 0; y < objTable.rows.length; y++) {
var checkbox = objTable.rows[y].childNodes[1].childNodes[0].childNodes[5];
 if (checkbox.checked == true) {
 if (num < 1) {
var renamebox =checkbox.parentNode.parentNode.parentNode.childNodes[3].childNodes[2];
var renametxt = checkbox.parentNode.parentNode.parentNode.childNodes[3].childNodes[0];
 renametxt.style.display = "none";
 renamebox.style.display = "block";
 num += 1;
}
 else { alert('不能同时重命名两个或两个以上文件');
    location.reload();
    break;
 }
   }
 }
}
```

（2）获取用户输入的文件名、文件路径，判断是否为文件夹，然后将数据传输至后台，实现重命名功能。

```
//确认重命名
 function sure() {
 var objTable = document.getElementById("tab");
 for (var y = 0; y < objTable.rows.length; y++) {
 var checkbox = objTable.rows[y].childNodes[1].childNodes[0].childNodes[5];
 if (checkbox.checked == true) {
 var changename= checkbox.parentNode.parentNode.parentNode
             .childNodes[3].childNodes[2].childNodes[1]
var name = $(changename).val();
var path = checkbox.parentNode.parentNode.parentNode.childNodes[5].innerHTML;
path = decodeURIComponent(path);
var imgpic = checkbox.parentNode.parentNode.parentNode
      .childNodes[3].getElementsByTagName("img");
if (imgpic[0].src.substring(imgpic[0].src.indexOf("assets"))=="assets/images/
Folder.png") {
 isDir = true;
 }
 else {
```

```
isDir = false;
}
 var data = {
        "path" : path,
        "name" : name,
        "isDir" : isDir
  };
    }
}
 $.ajax({
    url : "updatefile.action",
    type : "post",
    data : data,
    success : function(s) {
    if (s.success) {
    alert(s.msg);
    } else {alert(s.msg);}
    location.reload();
    }
});
}
```

### 5. 功能验证测试

项目部署到 Tomcat 中并发布，项目运行成功后，登录进入，选择文件夹进行重命名操作，如图 7-6 所示。

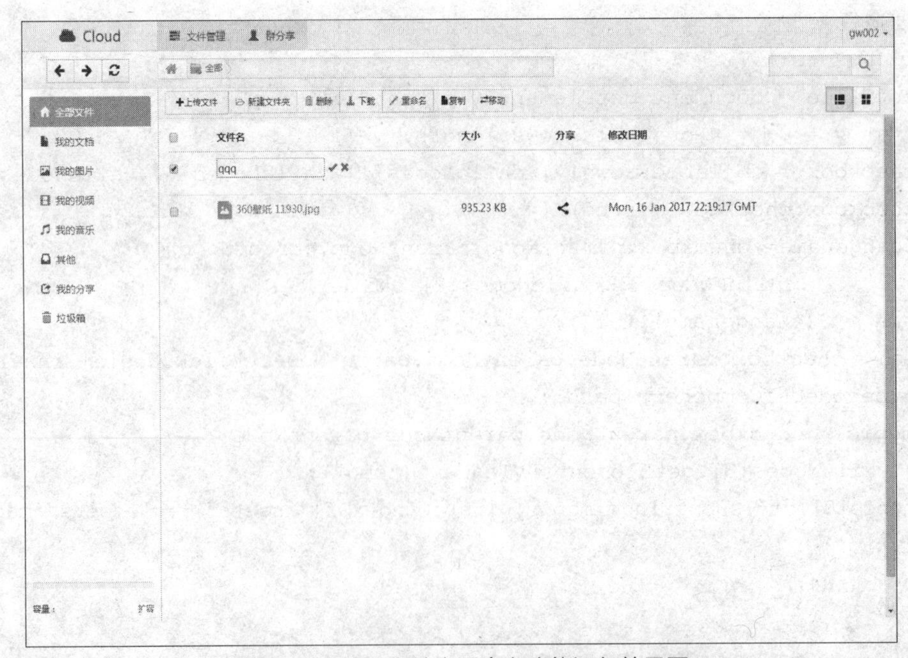


图 7-6　文件和文件夹重命名功能运行效果图

本功能的测试场景见表 7-3。

表 7-3　文件和文件夹重命名功能测试场景

| 编号 | 测试场景 | 输入参数 | 预期结果 |
|---|---|---|---|
| 1 | 文件的新名称不重复，不含特殊字符 | dirName：eeeee | 修改成功页面显示出来 |
| 2 | 文件的新名称重复，不含特殊字符 | dirName：eeeee | 提示创建失败，文件名已存在 |
| 3 | 文件的新名称不重复，含特殊字符 | dirName：eee/ee | 提示创建失败，文件名不合法 |
| 4 | 未选择文件单击重命名 | | 对话框不会弹出 |

## 任务 7.3　开发文件复制和粘贴功能

用户选中某一文件后，单击“复制”按钮，弹出复制对话框，选择目标目录，单击“确定”按钮，在目标目录下会出现复制的文件。

### 7.3.1　相关知识

本功能实现的技术原理如下。

（1）视图层：显示界面如图 7-7 所示。

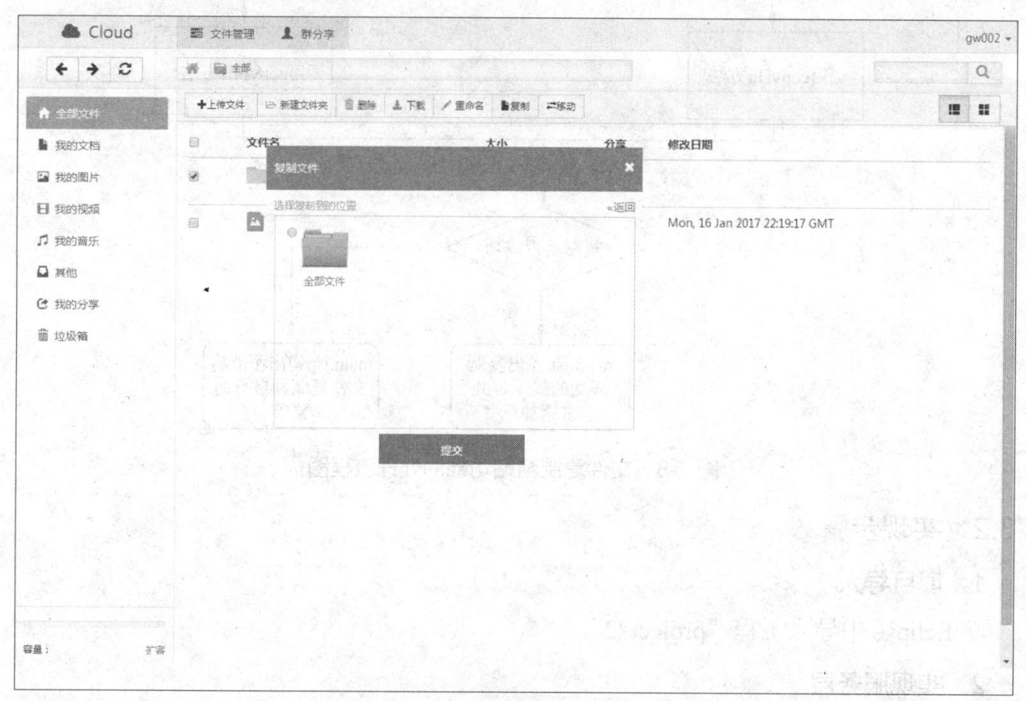


图 7-7　文件复制和粘贴操作界面

（2）控制层：接收视图层的消息，向服务层发送文件复制参数，并将服务层返回的结果发送到视图层。

（3）服务层：调用 Swift 服务对文件进行复制和粘贴，并将结果返回给控制层。

本功能的具体实现流程如图 7-8 所示。

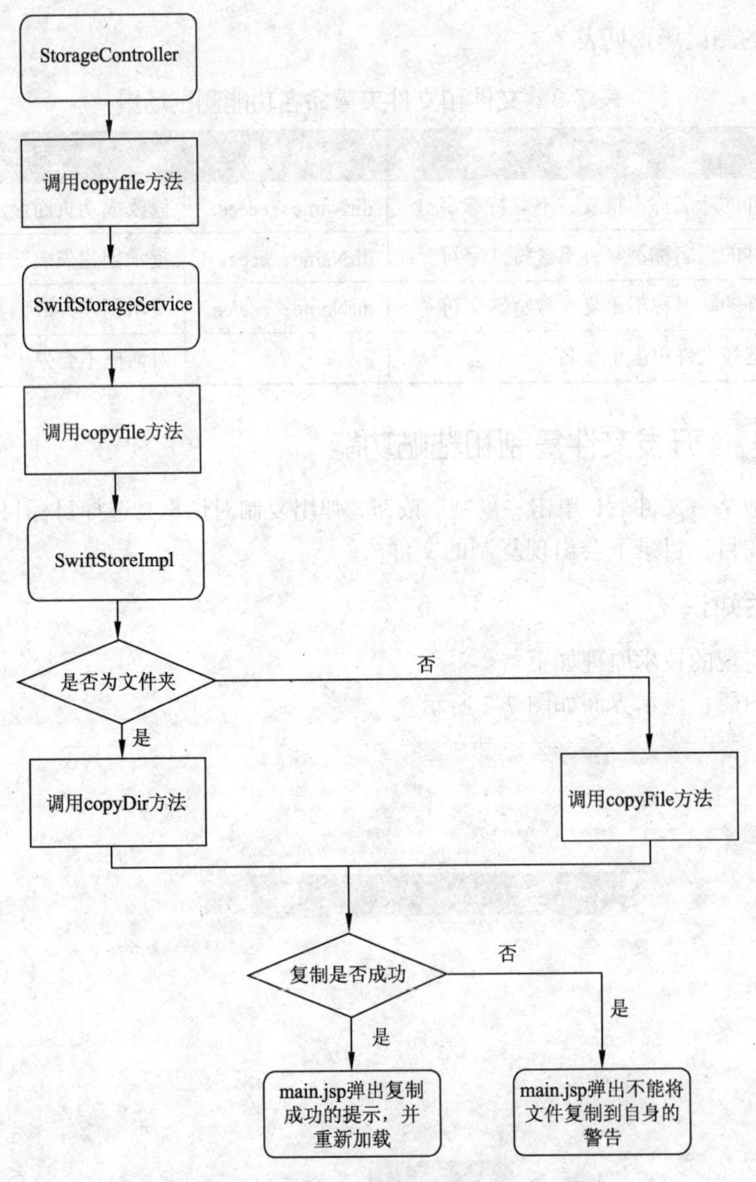


图 7-8　文件复制粘贴功能的处理流程图

### 7.3.2　实现步骤

#### 1. 项目导入

在 Eclipse 中导入工程“project72”。

#### 2. 实现服务层

（1）在“src/com/xiandian/cloud/storage/service/SwiftStorageService.java”中增加如下代码：

```
/**
     * 复制文件
```

```
     * @param username:用户名
     * @param password:密码
     * @param list:
     * @return
     */
    public boolean copyFile(String username, String password,List<Map<String,
Object>> list) {
        for (Map<String, Object> map : list) {
            String path = (String) map.get("path");
            String newpath = (String) map.get("newpath");
            String name = (String) map.get("name");
            String tisDir = (String) map.get("isDir");
            boolean isDir = "true".equals(tisDir) ? true : false;
            Account account = SwiftUtilTools.getAccount(username, password);
            SwiftStoreImpl Swiftdfs = new SwiftStoreImpl(account);
            if(isDir){
                Swiftdfs.copyDir(username + "/" + path, username + "/"+ newpath);
            }else{
                Swiftdfs.copyFile(username + "/" + path, username + "/"+ newpath
+ name);
            }
        }
        return true;
    }
    /**
     * 取得路径下的所有文件夹
     */
    public List getAllStoredDirectoryList(String username, String password,
            String path, String upath) {
        Account account = SwiftUtilTools.getAccount(username, password);
        SwiftStoreImpl Swiftdfs = new SwiftStoreImpl(account);
        List list = new ArrayList();
        String up = null;
        if (path == null || path.equals("")) {
            up = username;
        } else {
            up = username + "/" + path;
        }
        list = Swiftdfs.getDir(up);
```

```
        return list;
    }
```

（2）在“src/com/xiandian/cloud/storage/sh/SwiftStoreImpl.java”中增加如下代码：

```
/**
 * 描述：获取目录文件
 * @param rpath
 *            文件的目录
 * @return List
 */
public List getDir(String rpath) {
    Map mappath = getSplitPath(rpath);
    String path = mappath.get("path").toString();
    Container container = account.getContainer(mappath.get("rootPath")
            .toString());
    if (!path.contains("/")) {
        path = "/" + mappath.get("path").toString();
    }
    Collection<DirectoryOrObject> objects;
    if (path != null && path.length() > 1) {
        Directory directory = new Directory(path, '/');
        objects = listDirectory(container, directory);
    } else {
        objects = listDirectory(container);
    }
    List list = new ArrayList();
    for (DirectoryOrObject c : objects) {
        FileBean fb = new FileBean();
        fb.setName(c.getBareName());
        fb.setIsdirectory(c.isDirectory());
        if (c.isDirectory()) {
            String tpath = c.getAsDirectory().getName();
            fb.setPath(tpath);
            fb.setHaschild(haschild(container, tpath));
            list.add(fb);
        }

    }
    return list;
```

```
}
private boolean haschild(Container container, String path) {
    Collection<DirectoryOrObject> objects;
    if (path != null && path.length() > 1) {
        Directory directory = new Directory(path, '/');
        // objects = container.listDirectory(directory);
        objects = listDirectory(container, directory);
    } else {
        // objects = container.listDirectory();
        objects = listDirectory(container);
    }
    for (DirectoryOrObject c : objects) {
        if (c.isDirectory()) {
            return true;
        }
    }
    return false;
}
/**
 * 复制目录
 * @param rpath:要复制文件夹的目录
 * @param rnewpath:复制到某文件夹中的目录
 * @return
 */
public boolean copyDir(String rpath, String rnewpath) {
    Map mappath = getSplitPath(rpath);
    String rmpath = mappath.get("path").toString();
    String rootpath = mappath.get("rootPath").toString();
    Map rmappath = getSplitPath(rnewpath);
    String rnpath = rmappath.get("path").toString();
    int len = rootpath.length();
    String path = rmpath;
    String newpath = rnpath;
    String name = rmpath.split("/")[rmpath.split("/").length - 1];
    if (path.equals(newpath)) {
        return false;
    }
    copydir(rootpath, newpath + name + "/");
    copydirfile(len, rootpath, path, newpath, name, path);
```

```
    return true;
}

private void copydir(String rootpath, String newpath) {
    Container container = account.getContainer(rootpath);
    StoredObject newObject = getObject(container, newpath);
    if (newObject.exists()) {
        return;
    }
    uploadObject(newObject);
}

private void copydirfile(int len, String rootpath, String path,String newpath,
String name, String epath) {
    Container container = account.getContainer(rootpath);
    Directory directory = new Directory(epath, '/');
    Collection<DirectoryOrObject> objects = listDirectory(container,directory);
    for (DirectoryOrObject c : objects) {
        if (c.isDirectory()) {
            String cpath = c.getAsDirectory().getName();
            String tpath = cpath.substring(path.length());
            String tnewpath = newpath + name + "/" + tpath;
            copydir(rootpath, tnewpath);
            copydirfile(len, rootpath, path, newpath, name, cpath);
        } else {
            String temppath = c.getName();
            if (temppath.endsWith("/")) {
                continue;
            }
            String tpath = temppath.substring(path.length());
            String tnewpath = newpath + name + "/" + tpath;
            copyfile(rootpath, temppath, tnewpath);
        }
    }
}

/**
 * 复制文件
 * @param rpath:被复制的文件路径
```

```
 * @param rnewpath:新的文件路径
 * @return
 */
public boolean copyFile(String rpath, String rnewpath) {
    Map mappath = getSplitPath(rpath);
    String path = mappath.get("path").toString();
    if (!path.contains("/")) {
        path = mappath.get("path").toString();
    }
    Container container = account.getContainer(mappath.get("rootPath").toString());
    Map maprpath = getSplitPath(rnewpath);
    String rwpath = maprpath.get("path").toString();
    if (!rwpath.contains("/")) {
        rwpath = maprpath.get("path").toString();
    }
    StoredObject object = getObject(container, path);
    StoredObject newObject = getObject(container, rwpath);
    if (object.exists() && !newObject.exists()) {
        copyObject(object, container, newObject);
        return true;
    }
    return false;
}

private void copyfile(String rootpath, String path, String newpath) {
    Container container = account.getContainer(rootpath);
    StoredObject object = getObject(container, path);
    StoredObject newObject = getObject(container, newpath);
    copyObject(object, container, newObject);
}
```

3. 实现控制层

在“/src/com/xiandian/cloud/storage/Web/StorageController.java”中增加如下代码：

```
/**
    * 复制文件
    * @param request:请求
    * @param list:
    * @return
    */
```

```
    @RequestMapping("/copyfile")
    @ResponseBody
    public Object copyFile(HttpServletRequest request,@RequestBody List<Map
<String, Object>> list) {
        User user = getSessionUser(request);
        boolean flag = false;
        flag = storageService.copyFile(user.getUsername(),user.getPassword(),
list);
        return new MessageBean(flag, flag ? Constants.SUCCESS_8: Constants.
ERROR_9);
    }
    /**
     * 取得主页的文件夹
     *
     * @param request
     * @param response
     * @param path
     * @return
     */
    @RequestMapping("/homedir")
    @ResponseBody
    public Object homeDir(HttpServletRequest request,
            HttpServletResponse response, String path) {
        User user = getSessionUser(request);
        String upath = request.getSession().getServletContext()
                .getRealPath("/");
        List list = storageService.getAllStoredDirectoryList(user.getUsername(),
                user.getPassword(), path, upath);
        return new MessageBean(true, "", list);
    }
```

4. 实现前后台数据交互

当用户选中某一文件后，单击“复制”按钮，弹出复制对话框，用户选择复制的路径，然后单击“保存”按钮向后台发送相应的“json”数据完成复制。因此在前端实现的时候也分为以下两个步骤。

（1）获取选中文件，显示复制对话框。

① 用户选择某一文件时，触发单击事件“show()”，实现显示/隐藏“复制”按钮功能。

```
//显示重命名、复制选项
function show() {
```

```
 var objTable = document.getElementById("tab");
 for (var i = 0; i < objTable.rows.length; i++) {
 var checkbox = objTable.rows[i].childNodes[1].childNodes[0].childNodes[5];
 if (checkbox.checked == true) {
 document.getElementById("rename").style.display = "block";
 document.getElementById("copy").style.display = "block";
 break;
 }
 else {
document.getElementById("rename").style.display = "none";
document.getElementById("copy").style.display = "none";
}
}
}
```

② 实现复制对话框。

```
<!-- 复制 -->
 <div class="lb-copy" style="display: none;">
 <div class="container pl-266">
 <div class="move-box-head">
 <p>复制文件</p>
 <div class="lb-closeContainer">
 <a class="lb-close move-lb-close" href="javascript:void(0);">
 <span class="glyphicon glyphicon-remove"></span>
 </a>
 </div>
 </div>
 <div class="move-box">
 <div class="move-box-content">
 <div class="move-box-position">
 <div>
 <p class="move-position-head">选择复制到的位置</p>
 <div class="move-position-back">
<a href="javascript:void(0);" onclick="getdircopy()">&laquo;返回</a>
 </div>
 </div>
 <div class="move-copy-filder">
 <div class="myDiv"></div>
 </div>
```

```
</div>
<button class="btn btn-primary move-button" onclick="Copy();">提交</button>
</div>
</div>
</div>
</div>
```

③ 给“复制”按钮绑定单击事件“getdircopy()”，实现显示对话框，并动态显示根路径文件夹。

```
//对话框显示相应文件夹
 function getdircopy() {
 $(".lightboxOverlay").fadeIn();
 $(".lb-copy").fadeIn();
 var tmp = '<div class="col-md-3" style="margin-top: 10px;">'
     + '<input name="CheckS" type="radio" style="float:left" value="">'
     + ' <a href="javascript:void(0);" onclick="getdir()" >'
     + '<img src=assets/images/Folder_lg.png style="height:61px;width:61px">'
     +'<divstyle="height:21px;width:
87px;overflow:hidden;text-overflow:ellipsis;
      font-size:10pt;margin-left: 4px;">全部文件</div>'
     + '</a></div>';
 $(".myDiv").html(tmp);
}
```

④ 增加“getdir()”方法，获取该文件夹子目录下的所有文件夹。

```
function getdir(path) {
 var data = {path : path};
 $.ajax({
       url : "homedir.action",
       type : "post",
       data : data,
       success : function(s) {
 if (s.success) {
 var array = s.other;
 var tmp = '<div style="margin-top: 10px;">';
 for (var I = 0; I < array.length; i++) {
 if (array[i]['isdirectory'] == true) {
 if (array[i]['haschild'] == true) {
 tmp += ' <div class="col-md-3" style="margin-bottom: 20px;">'
    +'<input name="CheckS" type="radio" style="float:left" value=\''+array[i]
['path']+'\'>'
    +'<a  href="javascript:void(0);"  onclick="getdir(\''+  array[i]['path']+
'\')" >'
```

```
    +'<img src=assets/images/Folder_lg.png style="height:61px;width:61px">'
    +'<div style="height:21px;width: 87px;overflow:hidden;text-overflow:ellipsis;">'
    + array[i]['name']
    + '</div></a></div> ';
    } else {
tmp += ' <div class="col-md-3" style="margin-bottom: 20px;">'
+'<input  name="CheckS"  type="radio"  style="float:left"  value=\''+array[i]
['path']+'\'>'
    +'<img src=assets/images/Folder_lg.png style="height:61px;width:61px">'
+'<div style="height:21px;width: 84%;overflow:hidden;
text-overflow:ellipsis;margin-left: 13px;">'+ array[i]['name']+ '</div></div> ';
    }
  }
 }
 tmp += '</div>';
 $(".myDiv").html(tmp);
 }
    }
});
        }
```

（2）选择要复制到的文件夹目录，单击“保存”按钮实现复制。

① 为复制对话框“提交”按钮绑定单击事件“Copy()”。

```
<button class="btn btn-primary move-button" onclick="Copy();">提交</button>
```

② 获取文件原始路径，复制到新路径，判断选中的文件是否为文件夹及文件，然后将这些数据以json格式的形式传输到后台，实现复制功能。

```
//复制
 function Copy() {
  var objTable = document.getElementById("tab");
  var data = '[';
  var input1 = document.getElementsByName("CheckS");
  var newpath;
  for (var i = 0; i < input1.length; i++) {
//通过遍历，获取选中文件夹的路径
if (input1[i].checked == true) {
 newpath = $(input1[i]).val();
  }
 }
  if (newpath == undefined) {
alert("请选择复制位置");
return;
```

```
}
 for (var y = 0; y < objTable.rows.length; y++) {
var checkbox = objTable.rows[y].childNodes[1].childNodes[0].childNodes[5];
if (checkbox.checked == true) {
 var current_name = checkbox.parentNode.parentNode.parentNode
                        .childNodes[3].childNodes[2].childNodes[1]
 var name = $(current_name).val();
 var path = checkbox.parentNode.parentNode.parentNode.childNodes[5].innerHTML
 path = decodeURIComponent(path);
 var img2 = checkbox.parentNode.parentNode.parentNode.childNodes[3]
       .getElementsByTagName("img");
 if (img2[0].src.substring(img2[0].src.indexOf("assets"))=="assets/images/
Folder.png") {
isDir = true;
  }
  else {
 isDir = false;
 }
 data += '{"path":"' + path + '","newpath":"' + newpath+ '","isDir":"' + isDir
+ '","name":"' + name+ '"}';
 if (y < (objTable.rows.length - 1)) {
data += ',';
   }
  }
}
 data += ']';
 var data = eval('(' + data + ')');
 $.ajax({
    url : "copyfile.action",
    type : "post",
    contentType : "application/json; charset=utf-8",
    data : JSON.stringify(data),
    success : function(s) {
    if (s.success) {
              alert(s.msg)
    } else {
    alert(s.msg);
    }
    location.reload();
    }
```

```
});
}
```

5. 功能验证测试

将项目部署到 Tomcat 中并发布，项目运行成功后，登录进入，单击菜单栏的“复制”按钮，弹出复制对话框，运行效果如图 7-9 所示。

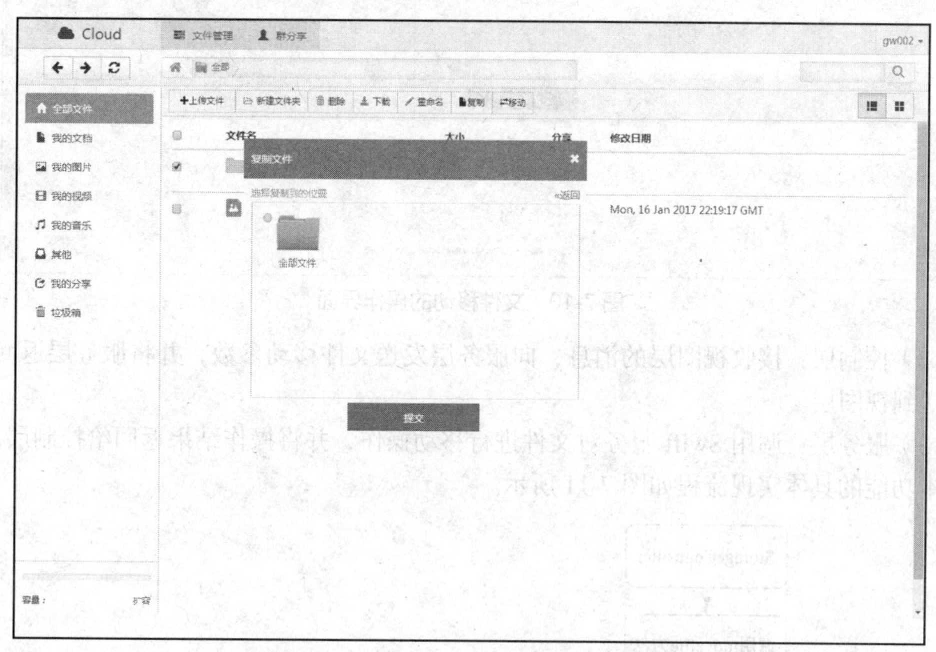


图 7-9　文件复制和粘贴功能运行效果图

本功能的测试场景见表 7-4。

表 7-4　文件复制和粘贴功能测试场景

| 编号 | 测试场景 | 输入参数 | 预期结果 |
|---|---|---|---|
| 1 | 选中文件后复制到其他目录 | | 文件在当前目录下显示 |
| 2 | 在当前文件目录下进行复制粘贴 | | 文件未做任何改变 |
| 3 | 复制文件夹 | | 提示不可以复制文件夹 |

## 任务 7.4　开发文件移动功能

用户选中某一文件后，单击“移动”按钮，弹出移动对话框，选择目标目录，单击“确定”按钮，将文件移动到在目标目录下。

### 7.4.1　相关知识

本功能实现的技术原理如下。

（1）视图层：显示界面如图 7-10 所示。

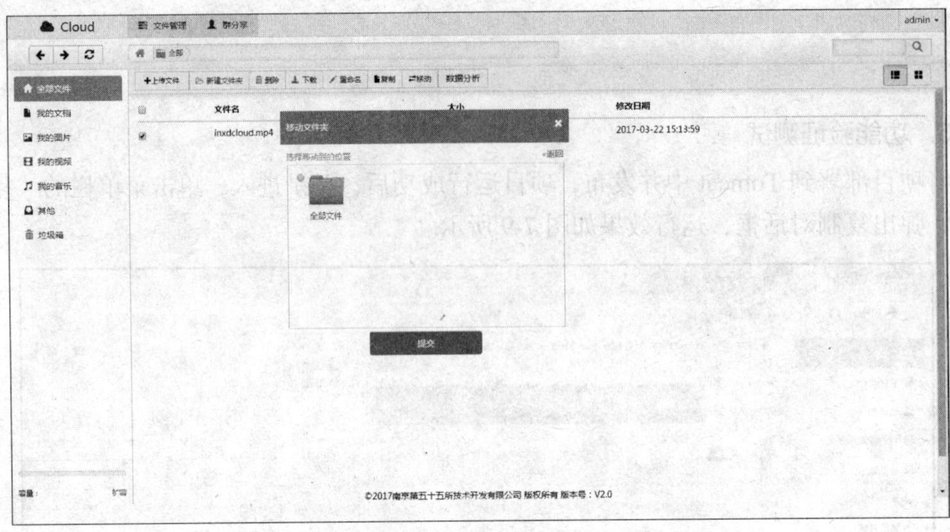


图 7-10　文件移动的操作界面

（2）控制层：接收视图层的消息，向服务层发送文件移动参数，并将服务层返回的结果发送到视图层。

（3）服务层：调用 Swift 服务对文件进行移动操作，并将操作结果返回给控制层。

本功能的具体实现流程如图 7-11 所示。

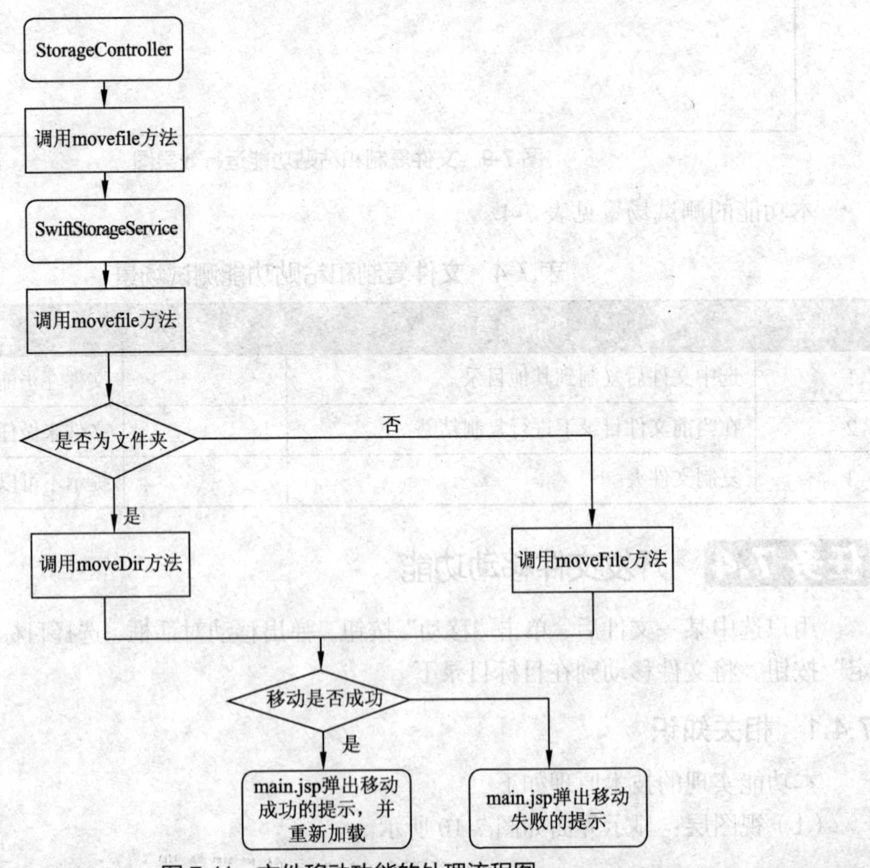


图 7-11　文件移动功能的处理流程图

### 7.4.2 实现步骤

#### 1. 项目导入

在Eclipse中导入工程“project73”。

#### 2. 实现服务层

（1）在“src/com/xiandian/cloud/storage/service/SwiftStorageService.java”中增加如下代码：

```
/**
 * 移动文件
 * @param username:用户名
 * @param password:密码
 * @param list:前台封装的json数组,形如[{path":"","newpath":"","isDir":"",
"name":""},{}]
 */
public boolean moveFile(String username, String password,List<Map<String,
Object>> list) {
    for (Map<String, Object> map : list) {
        String path = (String) map.get("path");
        String newpath = (String) map.get("newpath");
        String name = (String) map.get("name");
        String tisDir = (String) map.get("isDir");
        oolean isDir = "true".equals(tisDir) ? true : false;
        Account account = SwiftUtilTools.getAccount(username, password);
        SwiftStoreImpl Swiftdfs = new SwiftStoreImpl(account);
        if (isDir) {
            Swiftdfs.moveDir(username + "/" + path, username + "/"+ newpath);
        }else{
            Swiftdfs.moveFile(username + "/" + path, username + "/"+ newpath +
name);
        }
    }
    return true;
}
```

（2）在“src/com/xiandian/cloud/storage/sh/SwiftStoreImpl.java”中增加如下代码：

```
public boolean moveDir(String rpath, String rnewpath) {
    Map mappath = getSplitPath(rpath);
    String rmpath = mappath.get("path").toString();
    String rootpath = mappath.get("rootPath").toString();
```

```
    Map rmappath = getSplitPath(rnewpath);
    String rnpath = rmappath.get("path").toString();
    int len = rootpath.length();
    String path = rmpath;
    String newpath = rnpath;
    String name = rmpath.split("/")[rmpath.split("/").length - 1];
    if (path.equals(newpath)) {
        return false;
    }
    movedir(rootpath, path, newpath + name + "/");
    movedirfile(rootpath, path, newpath, name, path);
    return true;
}
private void movedir(String rootpath, String path, String newpath) {
    Container container = account.getContainer(rootpath);
    StoredObject object = getObject(container, path);
    StoredObject newObject = getObject(container, newpath);
    if (newObject.exists()) {
        return;
    }
    uploadObject(newObject);
    if (object.exists()) {
        delete(object);
    }
}
private void movedirfile(String rootpath, String path, String newpath,String
name, String epath) {
    Container container = account.getContainer(rootpath);
    Directory directory = new Directory(epath, '/');
    Collection<DirectoryOrObject> objects = listDirectory(container,directory);
    for (DirectoryOrObject c : objects) {
        if (c.isDirectory()) {
            String cpath = c.getAsDirectory().getName();
            String tpath = cpath.substring(path.length());
            String tnewpath = newpath + name + "/" + tpath;
            movedir(rootpath, cpath, tnewpath);
            movedirfile(rootpath, path, newpath, name, cpath);
        } else {
            String temppath = c.getName();
```

```
            if (temppath.endsWith("/")) {
                continue;
            }
            String tpath = temppath.substring(path.length());
            String tnewpath = newpath + name + "/" + tpath;
            movefile(rootpath, temppath, tnewpath);
        }
    }
}
private void movefile(String username, String path, String newpath) {
    Container container = account.getContainer(username);
    StoredObject object = getObject(container, path);
    StoredObject newObject = getObject(container, newpath);
    copyObject(object, container, newObject);
    delete(object);
}
/**
 * 移动文件
 * @param rpath: 准备要移动的文件路径
 * @param newpath:新的文件路径
 * @return
 */
public boolean moveFile(String rpath, String newpath) {
    Map mappath = getSplitPath(rpath);
    String path = mappath.get("path").toString();
    if (!path.contains("/")) {
        path = mappath.get("path").toString();
    }
    Container container = account.getContainer(mappath.get("rootPath").toString());
    Map newmappath = getSplitPath(newpath);
    String rnewpath = newmappath.get("path").toString();
    if (!rnewpath.contains("/")) {
        rnewpath = newmappath.get("path").toString();
    }
    StoredObject object = getObject(container, path);
    StoredObject newObject = getObject(container, rnewpath);
    if (object.exists() && !newObject.exists()) {
        copyObject(object, container, newObject);
        delete(object);
```

```
        return true;
    }
    return false;
}
```

3. 实现控制层

在“/src/com/xiandian/cloud/storage/Web/StorageController.java”中增加如下代码：

```
/**
 * 移动文件
 * @param request:请求
 * @param list:前台封装的json数组,形如[{path":"","newpath":"","isDir":"","name":""},
{}]
 */
@RequestMapping("/movefile")
@ResponseBody
public Object moveFile(HttpServletRequest request,@RequestBody List<Map<String,
Object>> list) {
    User user = getSessionUser(request);
    boolean flag = false;
    flag                                                                    =
storageService.moveFile(user.getUsername(),user.getPassword(),list);
    return  new  MessageBean(flag,  flag  ?  Constants.SUCCESS_9:  Constants.
ERROR_8);
}
```

4. 实现前后台数据交互

当用户选中某一文件后，单击“移动”按钮，弹出移动对话框，用户选择移动的路径，然后单击“保存”按钮向后台发送相应的“json”数据，完成移动功能，前端的实现也分为如下两个步骤。

（1）获取选中文件，显示移动对话框。

① 用户选择某一文件时，触发单击事件“show()”，为“show()”添加相应功能代码，实现显示/隐藏“移动”按钮。

```
//显示重命名、移动、复制选项
 function show() {
  var objTable = document.getElementById("tab");
  for (var i = 0; i < objTable.rows.length; i++) {
  var checkbox = objTable.rows[i].childNodes[1].childNodes[0].childNodes[5];
  if (checkbox.checked == true) {
  document.getElementById("rename").style.display = "block";
```

```
  document.getElementById("copy").style.display = "block";
document.getElementById("move").style.display = "block";
  break;
  }
  else {
document.getElementById("rename").style.display = "none";
document.getElementById("copy").style.display = "none";
}
}
}
```

② 实现移动对话框。

```
<!-- 移动 -->
<div class="lb-move" style="display: none;">
<div class="container pl-266">
<div class="move-box-head">
<p>移动文件夹</p>
<div class="lb-closeContainer">
<a class="lb-close move-lb-close" href="javascript:void(0);">
<span  class="glyphicon glyphicon-remove"></span>
</a></div></div>
<div class="move-box">
<div class="move-box-content">
<div class="move-box-position">
<div>
<p class="move-position-head">选择移动到的位置</p>
<div class="move-position-back">
<a href="javascript:void(0);" onclick="getdirmove()">&laquo;返回</a>
</div></div>
<div class="move-copy-filder">
<div class="myDiv"></div>
</div></div>
<button class="btn btn-primary move-button" onclick="Move();">提交</button>
</div>
</div>
</div>
</div>
```

③ 为“移动”按钮绑定单击事件“getmovedircopy ()”，显示对话框，并动态显示根路径文件夹。

```
//对话框显示相应文件夹
 function getmovedircopy() {
 $(".lightboxOverlay").fadeIn();
 $(".lb-copy").fadeIn();
 var tmp = '<div class="col-md-3" style="margin-top: 10px;">'
     + '<input name="CheckS" type="radio" style="float:left" value="">'
     + ' <a href="javascript:void(0);" onclick="getdir()" >'
     + '<img src=assets/images/Folder_lg.png style="height:61px;width:61px">'
     +'<divstyle="height:21px;width:
87px;overflow:hidden;text-overflow:ellipsis;
      font-size:10pt;margin-left: 4px;">全部文件</div>'
     + '</a></div>';
 $(".myDiv").html(tmp);
}
```

④ 增加"getdir()"方法，获取该文件夹子目录下的所有文件夹。

```
function getdir(path) {
 var data = {path : path};
 $.ajax({
        url : "homedir.action",
        type : "post",
        data : data,
        success : function(s) {
 if (s.success) {
 var array = s.other;
 var tmp = '<div style="margin-top: 10px;">';
 for (var i = 0; i < array.length; i++) {
 if (array[i]['isdirectory'] == true) {
 if (array[i]['haschild'] == true) {
 tmp += ' <div class="col-md-3" style="margin-bottom: 20px;">'
    +'<input name="CheckS" type="radio" style="float:left" value=\''+array[i]
['path']+'\'>'
    +'<a  href="javascript:void(0);"  onclick="getdir(\''+  array[i]['path']+
'\')" >'
    +'<img src=assets/images/Folder_lg.png style="height:61px;width:61px">'
   +'<div style="height:21px;width: 87px;overflow:hidden;text-overflow:ellipsis;">'
   + array[i]['name']
    + '</div></a></div> ';
   } else {
```

```
tmp += ' <div class="col-md-3" style="margin-bottom: 20px;">'
    +'<input name="CheckS" type="radio" style="float:left" value=\''+array[i]
['path']+'\'>'
    +'<img src=assets/images/Folder_lg.png style="height:61px;width:61px">'
    +'<div style="height:21px;width: 84%;overflow:hidden;
    text-overflow:ellipsis;margin-left: 13px;">'+ array[i]['name']+ '</div>
</div> ';
        }
    }
 }
 tmp += '</div>';
 $(".myDiv").html(tmp);
 }
    }
});
}
```

（2）选择要移动到的文件夹目录，单击“保存”按钮实现文件移动。

① 为移动对话框“提交”按钮绑定单击事件 Move()。

```
<button class="btn btn-primary move-button" onclick="Move();">提交</button>
```

② 获取文件原始路径，移动到新路径，判断选中的文件是文件夹还是文件名，然后将这些数据以“json”格式的形式传输到后台。

```
//移动
function Move() {
    var objTable = document.getElementById("tab");
    var data = '[';
    var input1 = document.getElementsByName("CheckS");
    var newpath;
    for (var i = 0; i < input1.length; i++) {
        if (input1[i].checked == true) {
            newpath = $(input1[i]).val();
        }
    }
    if (newpath == undefined) {
        alert("请选择移动位置");
        return;
    }
    for (var y = 0; y < objTable.rows.length; y++) {
        var checkbox = objTable.rows[y].childNodes[1].childNodes[0].childNodes[5];
```

```
        if (checkbox.checked == true) {
       var current_name = checkbox.parentNode.parentNode.parentNode.childNodes[3].
childNodes[2].childNodes[1];
            var name = $(current_name).val();
        var  path  =  checkbox.parentNode.parentNode.parentNode.childNodes[5].
innerHTML;
            path = decodeURIComponent(path);
            var img2 = checkbox.parentNode.parentNode.parentNode.childNodes[3]
                    .getElementsByTagName("img");
    if (img2[0].src.substring(img2[0].src.indexOf("assets"))=="assets/images/
Folder.png") {
                isDir = true;
            } else {
                isDir = false;
            }
            data += '{"path":"' + path + '","newpath":"' + newpath
                    + '","isDir":"' + isDir + '","name":"' + name
                    + '"}';
            if (y < (objTable.rows.length - 1)) {
                data += ',';
            }
        }
    }
    data += ']';
    var data = eval('(' + data + ')');
    $.ajax({
        url : "movefile.action",
        type : "post",
        contentType : "application/json; charset=utf-8",
        data : JSON.stringify(data),
        success : function(s) {
            if (s.success) {
                alert(s.msg);
            } else {
                alert(s.msg);
            }
            location.reload();
        }
    });
}
```

### 5. 功能验证测试

将项目部署到 Tomcat 中并发布，项目运行成功后，登录进入，单击菜单栏的“移动”按钮，弹出移动对话框，效果如图 7-12 所示。

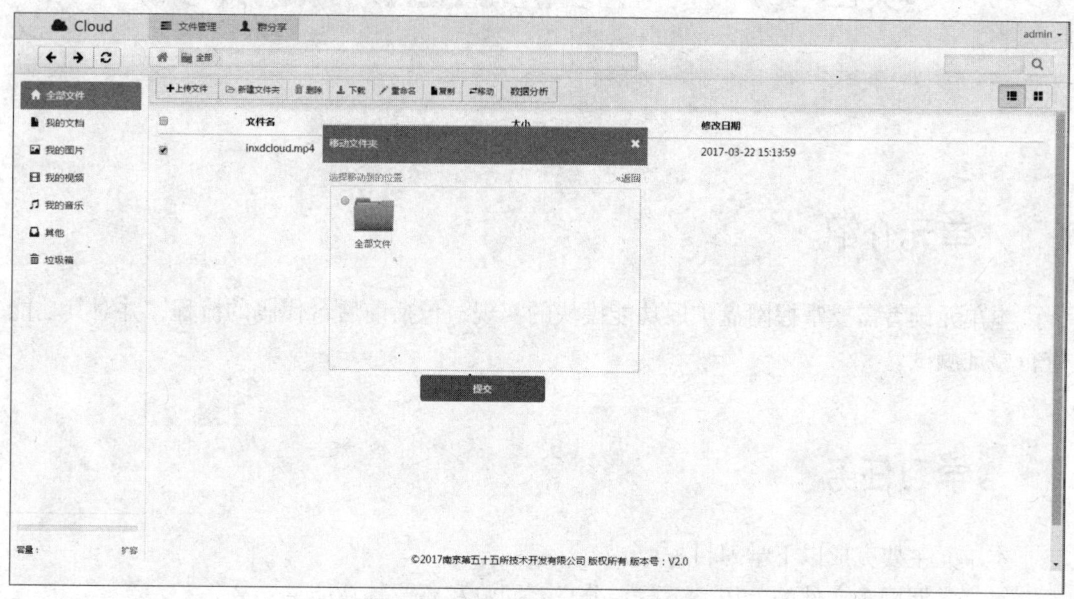


图 7-12　文件移动功能运行效果图

本功能的测试场景见表 7-5。

表 7-5　文件移动功能测试场景

| 编号 | 测试场景 | 输入参数 | 预期结果 |
|---|---|---|---|
| 1 | 选中文件后移动到其他目录 | | 文件在当前目录下显示 |
| 2 | 移动文件夹 | | 提示不可以移动文件夹 |

# 项目 8 开发功能扩展模块

## 单元介绍

本单元读者需要掌握网盘扩展功能模块的实现，包括前后台代码的流程，并对其功能进行验证测试。

## 学习任务

本单元主要完成以下学习目标：

- 掌握网盘文件的上传、下载、分享、回收和还原等功能。

根据这些学习目标，本项目将分解为以下 7 个任务，见表 8-1。

表 8-1 任务分解表

| 任务序号 | 任务内容 | 验证方式 |
|---|---|---|
| 任务 8.1 | 开发文件上传功能 | 可以从本地上传指定文件到网盘 |
| 任务 8.2 | 开发文件下载功能 | 可以从网盘下载指定文件到本地 |
| 任务 8.3 | 开发文件分享功能 | 可以将指定文件分享给其他用户 |
| 任务 8.4 | 开发群分享功能 | 可以将文件在群组内共享 |
| 任务 8.5 | 开发回收站功能 | 可以将文件删除到回收站 |
| 任务 8.6 | 开发清空回收站功能 | 可以将回收站的文件彻底删除 |
| 任务 8.7 | 开发还原文件功能 | 可以将回收站的文件恢复到原有位置 |

## 任务 8.1 开发文件上传功能

单击页面菜单中的“上传”按钮，弹出文件资源管理窗口，在文件资源管理器中选择所需上传的文件，单击“确定”按钮开始上传，上传结束后弹出提示框提示上传成功。

### 8.1.1 相关知识

本功能实现的技术原理如下。

（1）视图层：显示界面如图 8-1 所示。

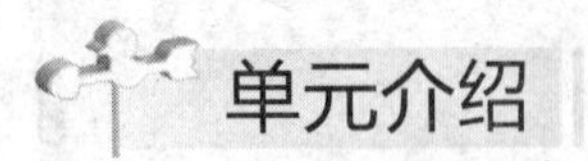

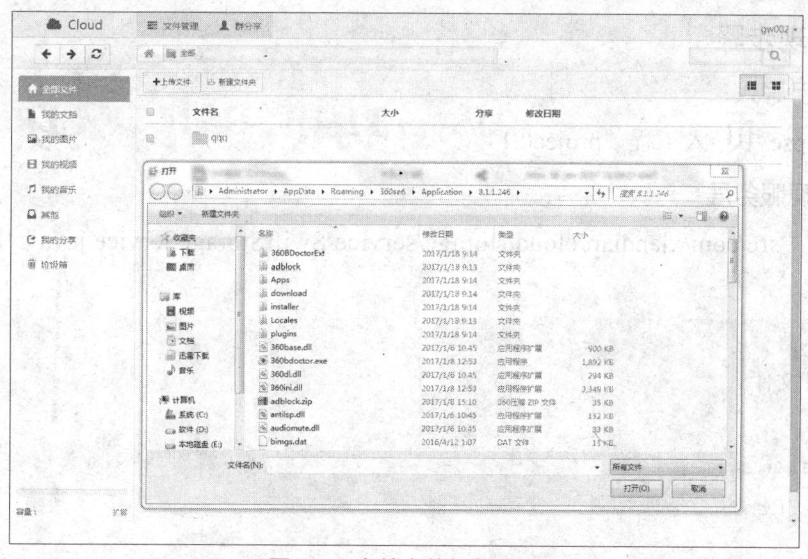

图 8-1　文件上传操作界面

（2）控制层：接收视图层的消息，向服务层发送文件上传参数，并将服务层返回的结果发送到视图层。

（3）服务层：调用 Swift 提供的上传接口实现文件上传，并将上传结果返回给控制层。

本功能的具体实现流程如图 8-2 所示。

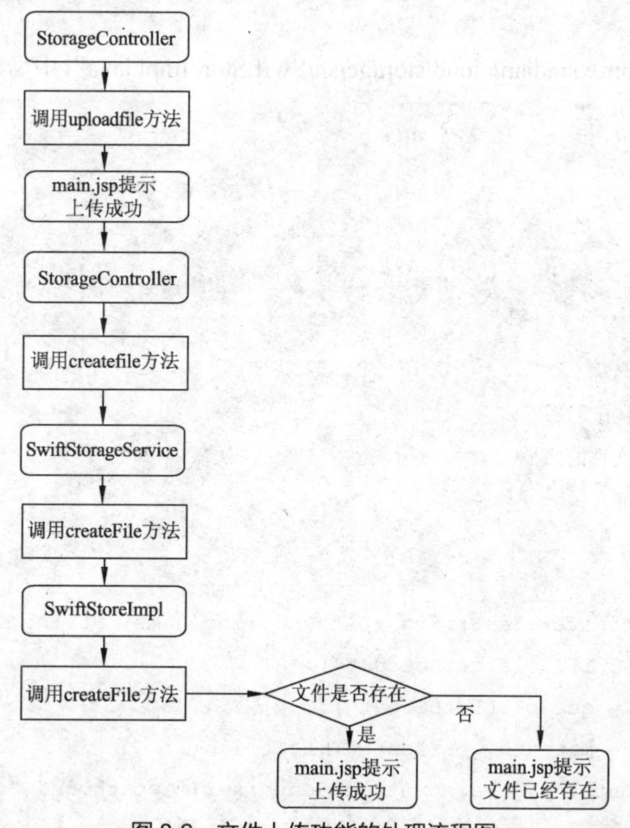


图 8-2　文件上传功能的处理流程图

### 8.1.2 实现步骤

1. 项目导入

在 Eclipse 中导入工程“project81”。

2. 实现服务层

（1）在“src/com/xiandian/cloud/storage/service/SwiftStorageService.java”中增加如下代码：

```
/**
     * 创建文件
     *
     * @param name
     * @return
     */
    public boolean createFile(String username, String password, String path,
            String name, String filepath) {
        Account account = SwiftUtilTools.getAccount(username, password);
        SwiftStoreImpl Swiftdfs = new SwiftStoreImpl(account);
        return Swiftdfs.createFile(username + "/" + path, name, filepath);
    }
```

（2）在“src/com/xiandian/cloud/storage/sh/SwiftStoreImpl.java”中增加如下代码：

```
/**
 * 描述：上传文件方法
 *
 * @param rpath
 *            当前目录
 * @param name
 *            文件名称
 * @param filepath
 *            上传文件的路径
 * @return
 *
 */
public boolean createFile(String rpath, String name, String filepath) {
    //account = UtilTools.getAccount();
    Map mappath = getSplitPath(rpath);
    String path = mappath.get("path").toString();
    Container container = account.getContainer(mappath.get("rootPath")
            .toString());
```

```
    // logger(mappath.get("rootPath").toString()+"执行 createFile 方法，创建文件");

    if (path == null || "".equals(path)) {
        // path = "/";
    }
    StoredObject object = getObject(container, path + name);
    boolean isexist = object.exists();
    if (isexist) {
        return false;
    }
    uploadObject(object, new File(filepath));
    return true;
}
private void uploadObject(StoredObject object, File file) {
    object.uploadObject(file);
}
```

### 3. 实现控制层

在“/src/com/xiandian/cloud/storage/Web/StorageController.java”中增加如下代码：

```
/**
 * 上传文件
 * @param request
 * @param file
 */
@RequestMapping("/uploadfile")
@ResponseBody
public  Object  uploadFile(HttpServletRequest  request,@RequestParam(value  =
"qqfile", required = true) MultipartFile file) {
    try {
        if (!file.isEmpty()) {
            byte[] bytes = file.getBytes();
        String upath = request.getSession().getServletContext().getRealPath("/");
            String uuid = UUID.randomUUID().toString();
            String path = "upload/" + uuid + file.getOriginalFilename();
            FileOutputStream fos = new FileOutputStream(upath + path);
            fos.write(bytes);
            fos.close();
            return new MessageBean(true, Constants.SUCCESS_4, path);
        } else {
```

```
            return new MessageBean(false, Constants.ERROR_6);
        }
    } catch (Exception e) {
        return new MessageBean(false, Constants.ERROR_6);
    }

    /**
     * 创建文件
     *
     * @param request
     * @param response
     * @param name
     */
    @RequestMapping("/createfile")
    @ResponseBody
    public Object createFile(HttpServletRequest request,
            HttpServletResponse response, String path, String name,
            String filepath) {
        User user = getSessionUser(request);
        String upath = request.getSession().getServletContext()
                .getRealPath("/");
        boolean flag = false;
        flag = storageService.createFile(user.getUsername(),user.getPassword(),
path, name, upath + filepath);
        return new MessageBean(flag, flag ? Constants.SUCCESS_4
                : Constants.ERROR_5);
    }
```

**4. 实现前后台数据交互**

这里采用“fine uploader”控件实现文件的上传，该上传控件有以下几个特点：

（1）支持文件上传进度显示；

（2）文件拖拽浏览器上传方式；

（3）Ajax 页面无刷新；

（4）多文件上传；

（5）跨浏览器；

（6）跨后台服务器端语言。

在使用该控件之前，先引入与该控件相关的 js 文件：

```
<script src="assets/javascripts/fineuploader.js"></script>
```

在 body 添加如下代码：

```
//上传文件
function createUploader() {
 var uploader = new qq.FineUploader({
  element : document.getElementById('result-uploader'),
  request : {endpoint : 'uploadfile.action'},            //上传文件的方法
  text : {
        uploadButton : '<i class="glyphicon glyphicon-plus"></i>上传文件'
        },
  template : '<div class="qq-uploader">'
        + '<pre class="qq-upload-drop-area"><span>{dragZoneText}</span></pre>'
        + '<div class="qq-upload-button btn btn-default btn-sm" style="background:
#F6F6F6;border: 1px solid rgb(206, 206, 206);display: inline;top: 3px;padding:
5px 15px 7px 15px;width: 100%;position: relative;overflow: hidden;border-radius:
0;">{uploadButtonText}</div>'
      + '<span class="qq-drop-processing" style="display:none"><span>
        {dropProcessingText}</span>'
        + '<span class="qq-drop-processing-spinner"></span></span>'
        + '<ul class="qq-upload-list" style="margin-top: 10px; text-align:
center;display:none"></ul>'
        + '</div>',
        classes : {
          success : 'alert alert-success',
          fail : 'alert alert-error'
        },
callbacks : {
        onComplete : function(id, fileName, responseJSON) {
        if (responseJSON.success) {
        var fielpath = responseJSON.other;
        createfile(fileName, fielpath);     //上传成功后在 hdfs 上生成最新文件
        }
      }
    }
});
}
 function createfile(fileName, filepath) {
    var data = {
        path : "${path}",
        name : fileName,
```

```
        filepath : filepath
    };
    $.ajax({
        url : "createfile.action",
        type : "post",
        data : data,
        success : function(s) {
        if(s.success){
            alert(s.msg);
            location.reload();
        }else{
            alert(s.msg);
            }
          }
    });
}
```

5. 功能验证测试

将项目部署到 Tomcat 中并发布，项目运行成功后，登录进入，单击菜单栏的“上传文件”按钮，选择需要上传的文件，如图 8-3 所示。

图 8-3 文件上传功能运行效果图

本功能的测试场景见表 8-2。

表 8-2　文件上传功能测试场景

| 编号 | 测试场景 | 输入参数 | 预期结果 |
|---|---|---|---|
| 1 | 登录成功 | 上传一个 TXT 文件 | 上传成功 |
| 2 | 登录成功 | 上传一个已存在的文件 | 上传失败，文件已存在 |
| 3 | 登录成功 | 上传一个无法识别的文件 | 上传失败，文件格式不对 |

## 任务 8.2　开发文件下载功能

单击所需下载文件的选择框，单击菜单栏中的“下载”按钮，文件通过浏览器下载到本地。如果下载的为文件夹，程序将文件夹中的文件打包为 ZIP 格式压缩包，再通过浏览器下载到本地。

### 8.2.1　相关知识

本功能实现的技术原理如下。

（1）视图层：显示文件下载的界面，如图 8-4 所示。

（2）控制层：接收视图层的消息，向服务层发送文件下载参数。

（3）服务层：调用 Swift 提供的下载接口实现文件下载。

本功能的具体实现流程如图 8-5 所示。

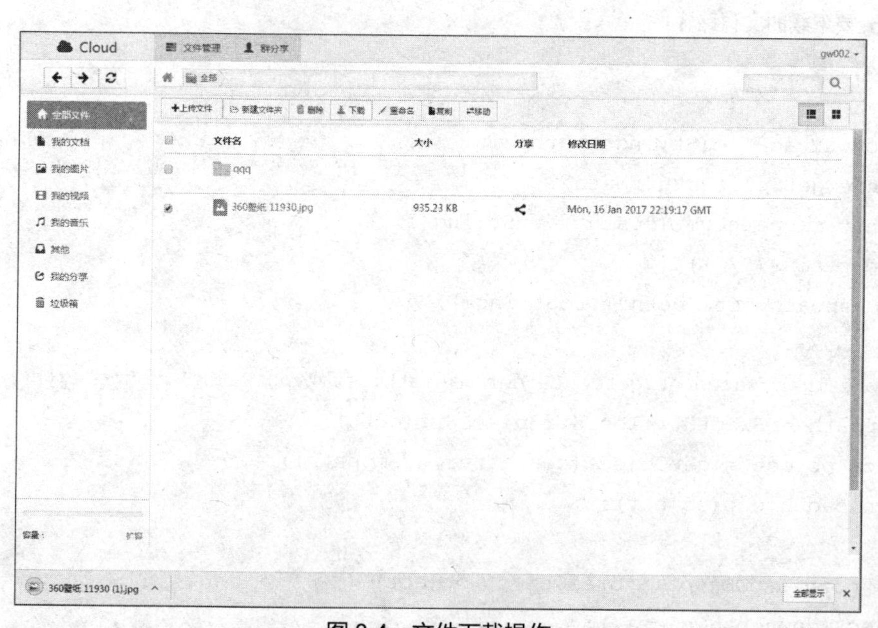


图 8-4　文件下载操作

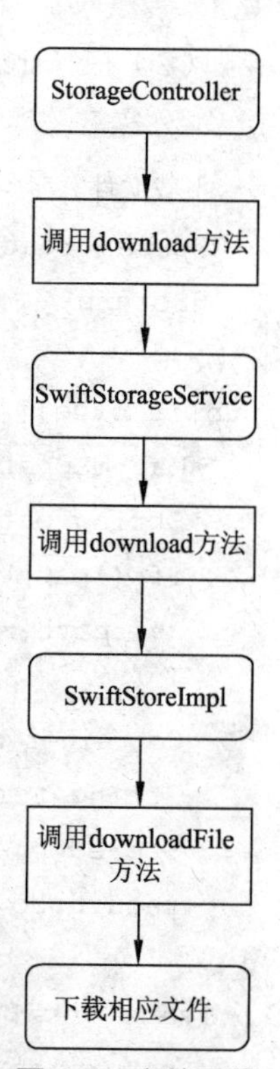


图 8-5　文件下载功能的处理流程图

### 8.2.2 实现步骤

#### 1. 项目导入

在 Eclipse 中导入工程“project82”。

#### 2. 实现服务层

（1）在“src/com/xiandian/cloud/storage/service/SwiftStorageService.java”中增加如下代码：

```
/**
 * 下载文件
 * @param username:用户名
 * @param password:密码
 * @param path:文件路径
 */
public byte[] download(String username, String password, String path) {
    Account account = SwiftUtilTools.getAccount(username, password);
    SwiftStoreImpl Swiftdfs = new SwiftStoreImpl(account);
    return Swiftdfs.downloadFile(username + "/" + path);
}
```

（2）在“src/com/xiandian/cloud/storage/sh/SwiftStoreImpl.java”类中增加如下代码：

```
/**
 * 下载文件
 * @param rpath 要下载的文件路径
 * @return
 */
public byte[] downloadFile(String rpath) {
    Map mappath = getSplitPath(rpath);
    String path = mappath.get("path").toString();
    if (!path.contains("/")) {
        path = mappath.get("path").toString();
    }
    Container container = account.getContainer(mappath.get("rootPath").toString());
    String temppath = SwiftUtilTools.replaceStr(path);
    StoredObject object = getObject(container, temppath);
    return downloadObject(object);
}
private byte[] downloadObject(StoredObject object) {
    return object.downloadObject();
}
```

3. 实现控制层

在“/src/com/xiandian/cloud/storage/Web/StorageController.java”中增加如下代码：

```
/**
 * 下载
 * @param request:请求
 * @param response:回应
 * @param paths:文件路径
 */
@RequestMapping("/download")
public void download(HttpServletRequest request,HttpServletResponse response,
String paths) {
    try {
        String[] strs = paths.split(",");
        int len = strs.length;
        if (len < 1) {
            return;
        }
        User user = getSessionUser(request);
        response.setContentType("text/html;charset=UTF-8");
        request.setCharacterEncoding("UTF-8");
        BufferedInputStream bis = null;
        BufferedOutputStream bos = null;
        String path = request.getSession().getServletContext().getRealPath("/");
        String ctxPath = path + "upload/temp";
        if (len == 1) {
            String filepath = strs[0];
            filepath = UtilTools.converStr(filepath);
            byte[] b = new byte[] {};
            b = storageService.download(user.getUsername(), user.getPassword(),
filepath);
            String[] strsss = filepath.split("/");
            String filename = strsss[strsss.length - 1];
            String downLoadPath = ctxPath + File.separator + filename;
            File downLoadFile = new File(downLoadPath);
            downLoadFile.createNewFile();
            BufferedOutputStream  buffer  =  new  BufferedOutputStream(new
FileOutputStream(downLoadFile));
            buffer.write(b);
```

```
            buffer.flush();
            buffer.close();
            long fileLength = downLoadFile.length();
            response.setContentType("application/octet-stream");
            response.setHeader("Content-disposition","attachment; filename="+
new String(filename.getBytes("utf-8"),"ISO8859-1"));
            response.setHeader("Content-Length", String.valueOf(fileLength));
            bis = new BufferedInputStream(new FileInputStream(downLoadPath));
            bos = new BufferedOutputStream(response.getOutputStream());
            byte[] buff = new byte[2048];
            int bytesRead;
            while (-1 != (bytesRead = bis.read(buff, 0, buff.length))) {
                bos.write(buff, 0, bytesRead);
            }
            bis.close();
            bos.close();
            downLoadFile.delete();
        } else {
            String[] strss = strs[0].split("/");
            String temps = strss[strss.length - 1];
            String[] strr = temps.split("\\.");
            strr[0] = UtilTools.converStr(strr[0]);
            String rootname = ctxPath + File.separator + strr[0];
            File rfile = new File(rootname);
            rfile.mkdir();
            for (int i = 0; i < len; i++) {
                String filepath = strs[i];
                filepath = UtilTools.converStr(filepath);
                byte[] b = new byte[] {};
                b = storageService.download(user.getUsername(), user.getPassword(),
filepath);
                String[] strsss = filepath.split("/");
                String filename = strsss[strsss.length - 1];
                String dpath = rootname + File.separator + filename;
                File dfile = new File(dpath);
                dfile.createNewFile();
                BufferedOutputStream  buff  =  new  BufferedOutputStream(new
FileOutputStream(dfile));
                buff.write(b);
```

```
            buff.flush();
            buff.close();
        }
        // 打包生成 tar.gz 文件
        File tarfile = new File(rootname + ".zip");
        tarfile.createNewFile();
        UtilTools.WriteToTarGzip(ctxPath + File.separator, strr[0],strr[0]
+ ".zip");
        // 下载 tar.gz 文件
        String downLoadPath = rootname + ".zip";
        String fp = "[批量下载]" + strr[0] + ".zip";
        File downLoadFile = new File(downLoadPath);
        long fileLength = downLoadFile.length();
        response.setContentType("application/octet-stream");
        response.setHeader("Content-disposition","attachment; filename="+
new String(fp.getBytes("utf-8"), "ISO8859-1"));
        response.setHeader("Content-Length", String.valueOf(fileLength));
        bis = new BufferedInputStream(new FileInputStream(downLoadPath));
        bos = new BufferedOutputStream(response.getOutputStream());
        byte[] buff = new byte[2048];
        int bytesRead;
        while (-1 != (bytesRead = bis.read(buff, 0, buff.length))) {
            bos.write(buff, 0, bytesRead);
        }
        bis.close();
        bos.close();
        downLoadFile.delete();
        UtilTools.deletefile(rootname);
    }
  } catch (Exception e) {
      e.printStackTrace();
  }
}
```

### 4. 实现前后台数据交互

当用户选中某一文件后，单击“下载”按钮将文件下载到本地，实现步骤如下。

（1）获取选中文件，显示“下载”按钮。

用户选择某一文件，触发单击事件“show()”，该事件显示“下载”按钮，代码如下：

```
function show() {
```

```
var objTable = document.getElementById("tab");
for (var i = 0; i < objTable.rows.length; i++) {
var checkbox = objTable.rows[i].childNodes[1].childNodes[0].childNodes[5];
if (checkbox.checked == true) {
document.getElementById("rename").style.display = "block";
document.getElementById("copy").style.display = "block";
document.getElementById("download").style.display = "block";//新增显示下载按钮
break;
}
else {
document.getElementById("rename").style.display = "none";
document.getElementById("copy").style.display = "none";
document.getElementById("download").style.display = "none";//新增隐藏下载按钮
 }
}
```

（2）为“下载”按钮绑定单击事件“downloadfile( )”，该事件获取用户选中文件的路径，将文件路径传向后台，实现下载功能，代码如下：

```
//下载文件
function downloadfile() {
 var objTable = document.getElementById("tab");
var data = '';
for (var y = 0; y < objTable.rows.length; y++) {
var checkbox = objTable.rows[y].childNodes[1].childNodes[0].childNodes[5];
 if (checkbox.checked == true) {
    var td3 = checkbox.parentNode.parentNode.parentNode.childNodes[3];
    var imgpic = td3.getElementsByTagName("img");
if (imgpic[0].src.substring(imgpic[0].src.indexOf("assets"))=="assets/images/Folder.png")
{
alert("文件夹中内容暂时无法提供下载，请进入文件夹选择相应文件进行下载");
return;
} //如果是文件夹则无法进行下载
else {
var path = checkbox.parentNode.parentNode.parentNode.childNodes[5].innerHTML;
path = decodeURIComponent(path);}
data += path + ",";
}
}
   if(data.length == 0){
       alert("请选择要操作的文件!!! ");
          return;
     };
```

```
    data = data.substring(0, data.length - 1);
    location.href = "download.action?paths=" + data;
}
```

5. 功能验证测试

将项目部署到 Tomcat 中并发布，项目运行成功后，登录进入，选择要下载的文件，勾选其复选框，单击菜单栏的“下载”按钮，开始下载文件，运行效果如图 8-6 所示。

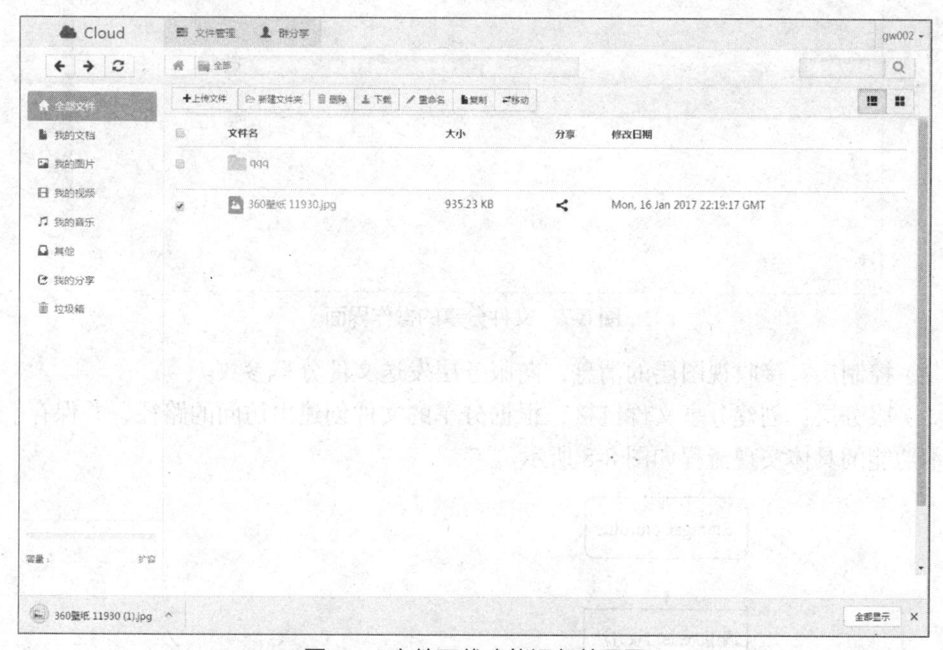


图 8-6 文件下载功能运行效果图

本功能的测试场景见表 8-3。

表 8-3 文件下载功能测试场景

| 编号 | 测试场景 | 输入参数 | 预期结果 |
| --- | --- | --- | --- |
| 1 | 登录成功后 | 选择一个文件 | 下载文件成功 |
| 2 | 登录成功后 | 选择多个文件 | 下载成功，为 ZIP 文件 |

## 任务 8.3 开发文件分享功能

在文件列表中每个文件都有一个“分享”按钮，单击即可分享当前文件。分享功能包含公开分享和私密分享，公开分享即任何人通过分享链接都可以获得文件，私密分享则还需要对应的密码才可以获得文件。

### 8.3.1 相关知识

本功能实现的技术原理如下。

（1）视图层：显示文件分享界面如图 8-7 所示。

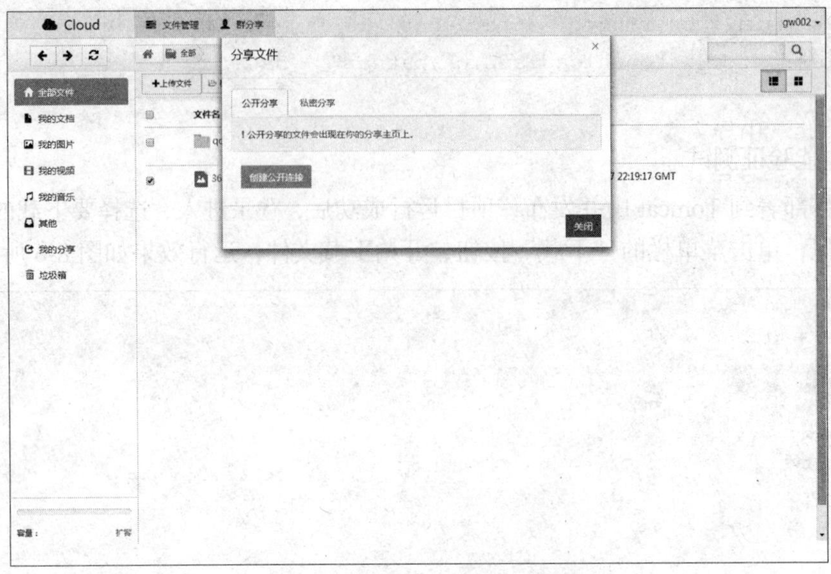


图 8-7　文件分享的操作界面

（2）控制层：接收视图层的消息，向服务层发送文件分享参数。

（3）服务层：创建分享文件链接，根据分享的文件创建出访问的路径，并保存。

本功能的具体实现流程如图 8-8 所示。

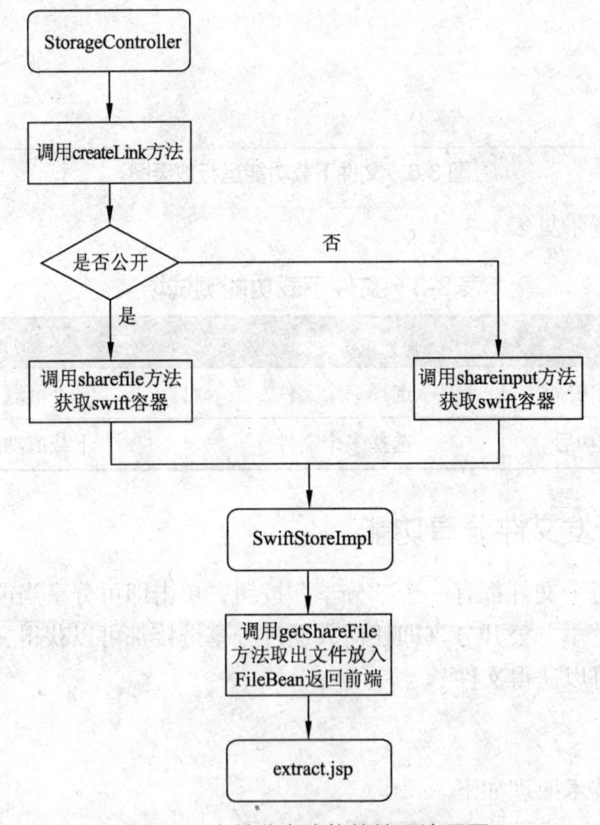


图 8-8　文件分享功能的处理流程图

## 8.3.2 实现步骤

### 1. 项目导入

在 Eclipse 中导入工程“project83”。

### 2. 实现服务层

在“src/com/xiandian/cloud/storage/sh/SwiftStoreImpl.java”中增加如下代码：

```
/**
 * 获取文件类表
 * @param username:用户名
 * @param path:当前目录
 * @return
 */
public List getShareFile(String username, String path) {
    Container container = account.getContainer(username);
    Collection<DirectoryOrObject> objects;
    List list = new ArrayList();
    if(path.contains("/")){
        Directory directory = new Directory(path, '/');
        objects = listDirectory(container, directory);
        for (DirectoryOrObject c : objects) {
            FileBean fb = new FileBean();
            fb.setName(c.getBareName());
            if (!c.isDirectory()) {
                StoredObject t = c.getAsObject();
                String tp = t.getPath();
                String ttp = SwiftUtilTools.replaceStr(tp);
                if(ttp.endsWith("/")) {
                    fb.setPath(t.getName());
                    fb.setIsdirectory(true);
                    list.add(fb);
                }
            }
        }
    }else{
        if (path != null && path.length() > 1) {
            Directory directory = new Directory(path, '/');
            objects = listDirectory(container, directory);
        } else {
            objects = listDirectory(container);
```

```
        }

        for (DirectoryOrObject c : objects) {
            FileBean fb = new FileBean();
            fb.setName(c.getBareName());
            fb.setIsdirectory(c.isDirectory());
            if (c.isDirectory()) {
                fb.setPath(c.getAsDirectory().getName());
                list.add(fb);
            } else {
                StoredObject t = c.getAsObject();
                String tp = t.getPath();
                String ttp = SwiftUtilTools.replaceStr(tp);

            if (!ttp.endsWith("/")) {
                    fb.setPath(container.getName()+"/"+t.getName());
                    fb.setLength(t.getContentLength());

    fb.setLastmodified(DateUtil.gmtToHMS(t.getLastModified()));
                    list.add(fb);
                }
            }
        }
    }
    return list;
}
```

### 3. 实现控制层

在“/src/com/xiandian/cloud/storage/Web/StorageController.java”中增加如下代码：

```
@RequestMapping("/createLink")
@ResponseBody
public Object createLink(HttpServletRequest request, int type,@RequestBody
List<Map<String, Object>> list) {
    int id = getSessionUser(request).getId();
    String scheme = request.getScheme();
    String serverName = request.getServerName();
    int serverPort = request.getServerPort();
    long currentTimeMillis = System.currentTimeMillis();
    String url = request.getContextPath();
```

```
    String port = serverPort == 80 ? "" : ":" + serverPort;
    String urlpath = scheme + "://" + serverName + port + url+ "/share.action?num="
+ currentTimeMillis;
    String identifyingCode = UtilTools.getCode();

    for (Map<String, Object> map : list) {
        String path = (String) map.get("path");
        String name = (String) map.get("name");
        String tisDir = (String) map.get("isDir");
        ShareBean shareBean = new ShareBean();
        shareBean.setUserid(id);
        shareBean.setIsdir(tisDir);
        shareBean.setFilename(name);
        shareBean.setFilepath(path);
        shareBean.setType(type);
        shareBean.setFilelength((String) map.get("filelength"));
        shareBean.setHttp(urlpath);
        shareBean.setDate(DateUtil.DateToString("yyyy-MM-dd    HH:mm:ss",new
Date()));
        if (type == 2) {
            shareBean.setPwd(identifyingCode);
        }
        shareService.saveShareFile(shareBean);
    }
    return new MessageBean(true, "",new ShareBean(urlpath, identifyingCode));
}
/**
 * 公开分享
 * @param request:请求
 * @param num:分享编号
 * @return
 */
@RequestMapping("/share")
public ModelAndView shareFile(HttpServletRequest request,String num) {
    User userExist = getSessionUser(request);
    String url = null;
    List<List> allList = new ArrayList();
    ModelAndView view = new ModelAndView();
    List<Object[]> unp = shareService.getUsernameBynum(num);
```

```
    String username = "";
    if(unp.size()>0){
        for(Object[] object:unp){
            int userid = (int) object[3];
            String pwd = (String) object[5];
            if (pwd != null) {
                if(userExist==null||userExist.getId()!=userid){
                    url = "/shareinput";
                }else{
                    url = "/extract";
                }
            } else {
                url = "/extract";
            }
            User user= userService.getUserByid(userid);
             username =user.getUsername();
            String rpath=object[2].toString();
    Account account = SwiftUtilTools.getAccount(username,user.getPassword());
            SwiftStoreImpl Swiftdfs = new SwiftStoreImpl(account);
            List lis=Swiftdfs.getShareFile(username,rpath);
            allList.add(lis);
        }
        view.setViewName(url);
        view.addObject("list", allList);
        view.addObject("num", num);
        view.addObject("shareusername", username);
        return view;
    }else{
        url = "/shareinput";
    // 设为-1 为 shareinput.jsp 展现"链接已失效或者资源不存在"
        view.addObject("num", "-1");
        view.setViewName(url);
        return view;
    }
}
/**
 * 私密分享
 * @param request:请求
 * @param num:分享编号
```

```
 * @param secruit:提取密码
 * @return
 */
@RequestMapping("/shareinput")
public ModelAndView shareInput(HttpServletRequest request,String num,String
secruit) {
    String url = null;
    List<List> allList = new ArrayList();
    ModelAndView view =new ModelAndView();
    List<Object[]> unp = shareService.getUsernameBynum(num);
    String username = "";
    if(unp.size()>0){
        for(Object[] object:unp){
            int userid = (int) object[3];
            String pwd = (String) object[5];
            if(secruit.equals(pwd)){
                User user= userService.getUserByid(userid);
                username =user.getUsername();
                String rpath=object[2].toString();
    Account account = SwiftUtilTools.getAccount(username,user.getPassword());
                SwiftStoreImpl Swiftdfs = new SwiftStoreImpl(account);
                List lis=Swiftdfs.getShareFile(username, rpath);
                allList.add(lis);
                url = "/extract";
            } else {
                url = "/shareinput";
            view.addObject("msg", "<font color='red'>验证错误!! </font>");
            }
        }
        view.setViewName(url);
        view.addObject("num", num);
        view.addObject("list", allList);
        view.addObject("shareusername", username);
    }
    return view;
}
```

4．实现前后台数据交互

当用户选中某一文件后，单击“分享”按钮，弹出分享对话框，用户可以选择公开或私密分享，然后单击“创建分享链接”按钮向后台发送请求，经过后台处理并生成一个分享链接返回至前端，完成分享。因此在前端实现的时候也分为以下两个步骤。

（1）获取选中文件，显示复制对话框。

① 用户选择某一文件时，触发单击事件“show()”，为事件“show()”添加相应功能代码，实现显示/隐藏分享按钮功能。

```
function show() {
 var objTable = document.getElementById("tab");
 for (var i = 0; i < objTable.rows.length; i++) {
 var checkbox = objTable.rows[i].childNodes[1].childNodes[0].childNodes[5];
 if (checkbox.checked == true) {
 document.getElementById("rename").style.display = "block";
 document.getElementById("copy").style.display = "block";
 document.getElementById("download").style.display = "block";
 document.getElementById("share").style.display = "block";
 break;
}
 else {
 document.getElementById("rename").style.display = "none";
 document.getElementById("copy").style.display = "none";
 document.getElementById("download").style.display = "none";
 document.getElementById("share").style.display = "none";
 }
}
```

② 实现文件分享对话框。

```
<!-- 分享 -->
<div id="modal3" class="modal1 mymodal" aria-hidden="true"
 style="display: none;">
<div class="modal-dialog ">
<div class="modal-content">
<div class="modal-header">
<button type="button" class="close gb" data-dismiss="modal"
aria-hidden="true" onclick="close1()">×</button>
<h4 class="modal-title1">分享文件</h4></div>
<div class="modal-body">
<div class="tab-v1">
<ul class="nav nav-tabs">
<li class="active"><a href="#home1" data-toggle="tab"
onclick="allshare()">公开分享</a></li>
```

```
<li class=""><a href="#profile" data-toggle="tab"
onclick="secshare()">私密分享</a></li>
</ul>
<div class="tab-content">
<div id="home1" class="tab-pane fade active in">
<div class="row" style="margin-left: -1px; margin-right: -1px">
<div class="alert alert-warning ">
公开分享的文件会出现在你的分享主页，其他人都能查看下载
</div>
<div class="col-md-8" id="publicLink">
<p>
<button class="btn rounded btn-info" type="button" onclick="createlink(1)">
 创建公开链接
 </button>
 </p>
 </div>
 </div>
 </div>
<div id="profile" class="tab-pane fade">
<div class="row" style="margin-left: -1px; margin-right: -1px">
<div class="alert alert-warning ">只有分享的好友能看到，其他人看不到</div>
<div class="col-md-8" id="securitLink">
 <p>
 <button class="btn rounded btn-info" type="button" onclick="createlink(2)">
    <i class="glyphicon glyphicon-lock"></i>创建私密链接</button>
                                        </p>
                                    </div>
                                </div>
                            </div>
                        </div>
                    </div>
                </div>
                <div class="modal-footer">
                    <button  type="button"  class="btn  btn-primary"  onclick=
"close1()">关闭</button>
                </div>
            </div>
        </div>
    </div>
```

③ 公开分享/私密分享切换显示。

```
//公开分享显示
function allshare(){
 $("#home1").addClass('active');
 $("#home1").addClass('in');
 $("#profile").removeClass('active');
 $("#profile").removeClass('in');
}
//私密分享显示
function secshare(){
 $("#home1").removeClass('active');
 $("#home1").removeClass('in');
 $("#profile").addClass('active');
 $("#profile").addClass('in');
}
```

（2）为“分享”按钮绑定单击事件“share()”，实现对话框的显示，并获取分享文件的路径、是否是文件夹、文件名称、文件长度等信息。

```
function share() {
$("#modal3").css("display", "block");
var objTable = document.getElementById("tab");
var data = '[';
for (var y = 0; y < objTable.rows.length; y++) {
var checkbox = objTable.rows[y].childNodes[1].childNodes[0].childNodes[5];
if (checkbox.checked == true) {
var current_name = checkbox.parentNode.parentNode
            .parentNode.childNodes[3].childNodes[2].childNodes[1];
var name = $(current_name).val();
var filelength = checkbox.parentNode
            .parentNode.parentNode.childNodes[7].innerHTML;
var path = checkbox.parentNode.parentNode
            .parentNode.childNodes[5].innerHTML;
path = decodeURIComponent(path);
var img2 = checkbox.parentNode.parentNode.parentNode.childNodes[3]
                                   .getElementsByTagName("img");
if
(img2[0].src.substring(img2[0].src.indexOf("assets"))=="assets/images/Folder
.png"){
    isDir = true;
}
else {
    isDir = false;
}
```

```
data += '{"path":"' + path + '","isDir":"' + isDir + '","name":"' + name +
'","filelength":"' + filelength + '"}';
if (y < (objTable.rows.length - 1)) {
 data += ',';
 }
}
}
data += ']';
var data1 = eval('(' + data + ')');
return data1;
}
```

选择分享文件的方式创建相应的分享链接。

```
//创建分享链接，参数 type（1 为共享链接，2 为私密链接）
function createlink(type){
 var data = share();    //获取相应数据
 $.ajax({
    url:"createLink.action?type="+type,
    type:'post',
    contentType : "application/json; charset=utf-8",
    data : JSON.stringify(data),
    success:function(s){
 if(s.success){
    if(type=='1'){
    var container='<div class="panel-heading">'+'</div>'
    +'<div class="panel-body" style=" width: 524px;"> '
    +'<div class="form-horizontal" role="form">'
    +'<div class="form-group">'
    +' <label for="1" style=" margin-left: -44px;" class="col-lg-2 control-
label">链接
    </label>'
    +'<div class="col-lg-10">'
    +'  <input  class="form-control"  id="1"  value="'+s.other.http+'"  type=
"text">'
    +' </div>' +' </div>'+' <div class="form-group">'
    +' <div class="col-lg-offset-2 col-lg-10" style="text-align:right">'
    +'<button type="submit" class="btn-u btn-u-green"
                 onclick="copyToClipBoard(3,2,1)" id="3">
     复制</button>'
     +' </div>'+' </div>' +' </div>' +' </div>';
     $("#publicLink").empty().append(container);
     copyToClipBoard(3,2,1);
      }
    else{
```

```
    var container='<div class="panel-heading">'+'</div>'
    +'<div class="panel-body" style=" width: 524px;"> '
    +'<div class="form-horizontal" role="form">'
    +'<div class="form-group">'
    +' <label for="4" style="margin-left: -44px;" class="col-lg-2 control-
label">
       链接</label>'
    +'<div class="col-lg-10">'
    +' <input class="form-control" id="4" value="'+s.other.http+'" type="text">'
    +' </div>'+' </div>'+'<div class="form-group">'
    +'<label for="inputPassword1" style="margin-left: -44px;" class="col-lg-2
                                  control-label">密码</label>'
    +' <div class="col-lg-4">'
    +'<input class="form-control" id="5" value="'+s.other.pwd+'" type="text">'
    +'</div>'+' </div>' +' <div class="form-group">'
    +' <div class="col-lg-offset-2 col-lg-10" style="text-align:right">'
    +'<button type="submit" class="btn-u btn-u-green" id="6"
                  onclick="copyToClipBoard(6,5,4)">复制</button>'
    +' </div>'+' </div>'+' </div>'+' </div>';
    $("#securitLink").empty().append(container);
    copyToClipBoard(6,5,4);
    }
    }
    }
    });
}
//私密分享显示密码
function copyToClipBoard(a,b,c){
 var id='#'+a;
 $(id).zclip({
     path: "assets/javascripts/ZeroClipboard.swf",
     copy: function(){
     var secruit='';
     if(a=='6'){secruit+='密码: '+$('#'+b).val();}
     return $('#'+c).val()+'   '+secruit;
    }
    });
}
```

5. 功能验证测试

将项目部署到 Tomcat 中并发布，项目运行成功后，登录进入，单击要分享的文件右侧的“分享”按钮，选择分享类型，取得分享的链接，如图 8-9 所示。

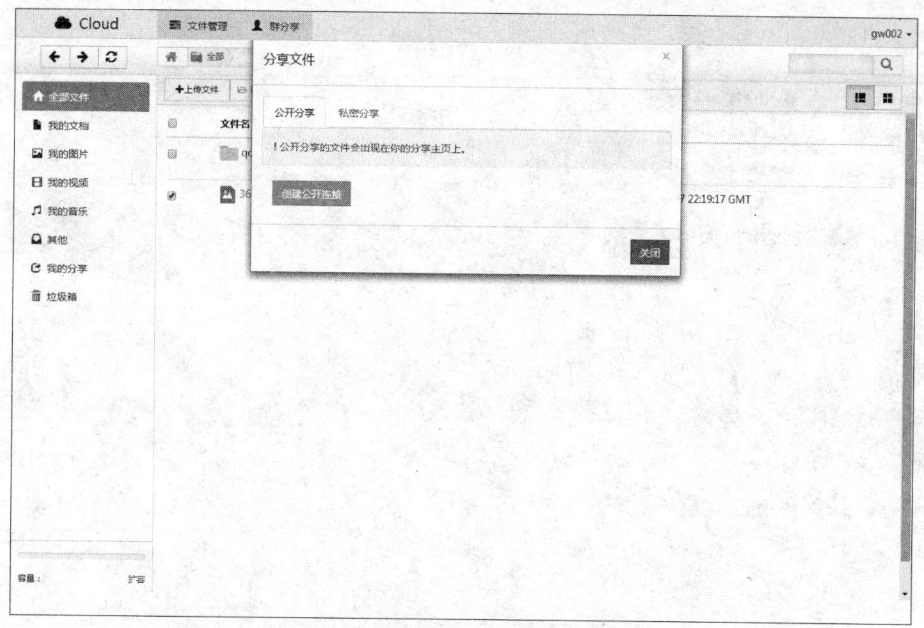


图 8-9　文件分享功能运行效果图

本功能的测试场景见表 8-4。

表 8-4　文件分享功能测试场景

| 编号 | 测试场景 | 输入参数 | 预期结果 |
|---|---|---|---|
| 1 | 单击文件的分享按钮 | 选择公开分享，创建文件分享链接 | 通过链接可以访问分享的文件 |
| 2 | 单击文件的分享按钮 | 选择私密分享，创建文件分享链接和密码 | 访问链接输入密码后得到分享的文件 |
| 3 | 单击文件夹的分享按钮 | 选择公开分享，创建文件夹分享链接 | 通过链接可以访问分享文件夹的全部内容 |
| 4 | 单击文件夹的分享按钮 | 选择私密分享，创建文件夹分享链接和密码 | 访问链接输入密码后得到分享文件夹的全部内容 |

## 任务 8.4　开发群分享功能

在群分享列表中，每个群组中都有一个公共的文件管理列表，创建群组的人能邀请用户共同管理群组中的文件。

### 8.4.1　相关知识

本功能实现的技术原理如下。

（1）视图层：显示群分享的界面，如图 8-10 所示。

（2）控制层：接收视图层的消息，向服务层发送群分享参数，并将服务层返回的结果发送到视图层。

（3）服务层：调用 Swift 接口创建群组、邀请用户、管理文件，并将处理结果返回给控制层。

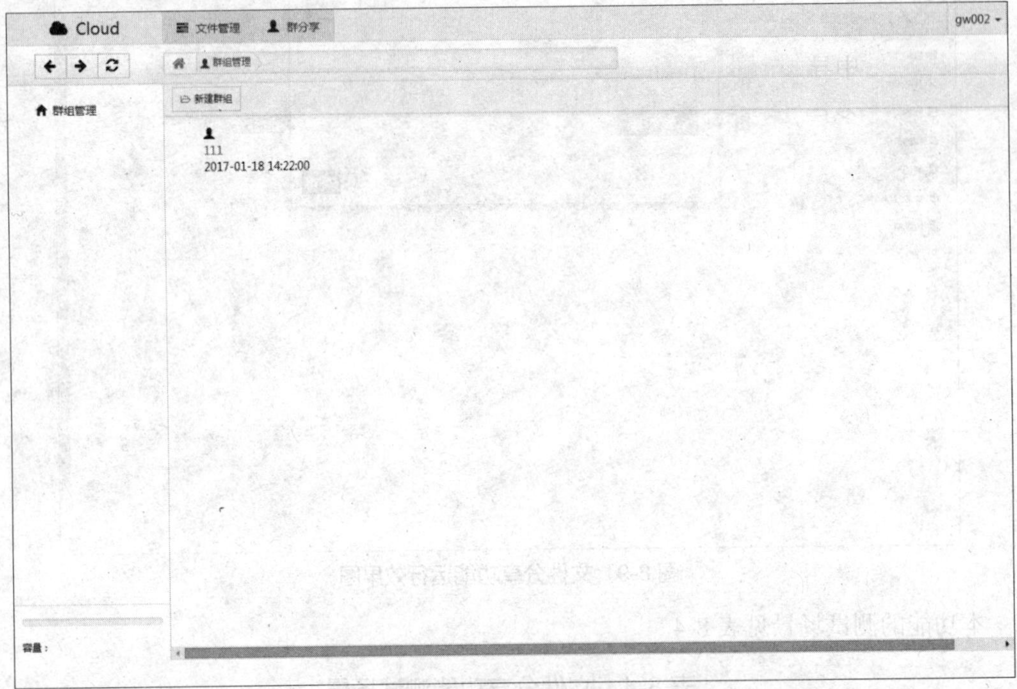


图 8-10　群分享操作界面

本功能的具体实现流程如图 8-11 所示。

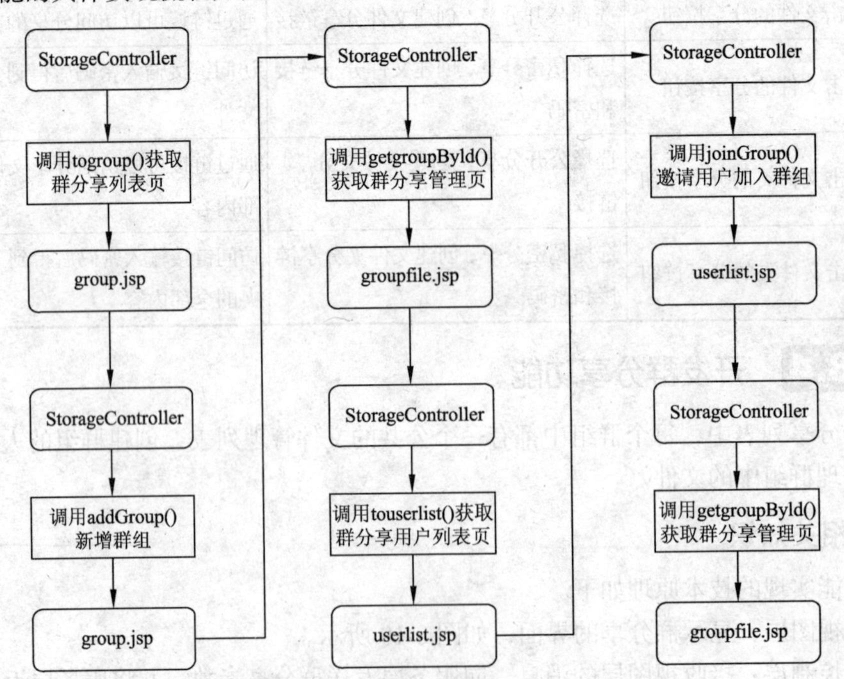


图 8-11　群分享功能的处理流程图

### 8.4.2 实现步骤

1. 项目导入

在 Eclipse 中导入工程“project84”。

2. 实现服务层

```
/**
 * 获取所有群组
 *
 * @return
 */
public List<Group> getall(){
    return groupDao.getAll();
}

/**
 * 根据用户 id 查询群组
 *
 * @param userid 用户 id
 * @return
 */
public List<Group> getGroup(int userid){
    return groupDao.getGroup(userid);
}

/**
 * 添加群组
 *
 * @param userid 用户 id
 * @param name  群组名称
 * @param date  添加日期
 */
public void addGroup(int userid,String name,String date){
    groupDao.addGroup(userid, name, date);
    SwiftDFS swiftDFS=new SwiftDFS();
    swiftDFS.createContainer("group_"+name);
}
```

3. 实现控制层

在“/src/com/xiandian/cloud/storage/Web/StorageController.java”中增加如下代码：

```
/**
 * 群分享列表页
 * @param request:请求
 * @return
 */
@RequestMapping("/togroup")
public ModelAndView toGroup(HttpServletRequest request) {
    User user = getSessionUser(request);
    int userid = user.getId();
    List<Object> groupid = groupService.findGroupByUserId(userid);
    List list = groupService.getGroup(userid);
    String url = "/group";
    ModelAndView view = new ModelAndView(url);
    view.addObject("list", list);
    return view;
}
/**
 * 新建群组
 * @param request:请求
 * @param name:群组名
 * @return
 */
@RequestMapping("/addGroup")
@ResponseBody
public Object addGroup(HttpServletRequest request, String name) {
    User user = getSessionUser(request);
    groupService.addGroup(user.getId(),
name,DateUtil.DateToString("yyyy-MM-dd HH:mm:ss", new Date()));
    return true;
}
/**
 * 群分享用户列表
 * @param request
 * @param id
 * @return
 */
@RequestMapping("/touserlist")
public ModelAndView touserlist(HttpServletRequest request,int id) {
    User user = getSessionUser(request);
```

```
    List<Object[]> groupMember = groupService.findGroupMember(id);
    ArrayList lists = new ArrayList();
    for (Object[] obj : groupMember) {
        User user1 = new User();

        user1.setId(Integer.parseInt(obj[1].toString()));
        user1.setUsername(obj[0].toString());
        lists.add(user1);
    }
    List<User> list = userService.getAll();
    String url = "/userlist";
    ModelAndView view = new ModelAndView(url);
    view.addObject("list", list);
    view.addObject("groupid", id);
    view.addObject("groupMember", lists);
    return view;
}
/**
 * 邀请用户加入群组
 * @param groupid
 * @param usersid
 * @return
 */
@RequestMapping("/joinGroup")
@ResponseBody
public Object joinGroup(int groupid, int usersid) {
    groupService.joinGroup(groupid, usersid);
    return true;
}
/**
 * 群分享管理页
 * @param request:请求
 * @param id:群组 id
 * @return
 */
@RequestMapping("/getgroupById")
public ModelAndView getGroupById(HttpServletRequest request, int id) {
    User user = getSessionUser(request);
    List group = groupService.getGroupById(id);
```

```
    String path = Constants.RESOURCE_FILE + Constants.GROUP_FILE + "/" + id;
   List list = storageService.getAllStoredList(user.getUsername(),user.getPassword(),
path);
    String url = "/groupfile";
    ModelAndView view = new ModelAndView(url);
    view.addObject("id", id);
    view.addObject("path", path);
    view.addObject("group", group);
    view.addObject("list", list);
    return view;
}
```

4. 实现前后台数据交互

(1) 获取群组列表。

```
<c:forEach var="group" items="${list}">
<div class=" col-lg-3 col-md-4 col-sm-6 col-xs-12 gridview-imgbox">
<i class="glyphicon glyphicon-user"></i>
<div class="text">
<div class="content">
<div><a href="getgroupById.action?id=${group.id}">${group.name}</a></div>
<div>${group.date}</div>
</div>
</div>
</div>
</c:forEach>
```

(2) 实现群组。

① 实现群组对话框。

```
<div   id="groupform"   class="modal   fade"   aria-labelledby="myModalLabel"
aria-hidden="true" tabindex="-1" role="dialog" >
<div class="modal-dialog ">
<div class="modal-content">
<div class="modal-header">
<button type="button" class="close" data-dismiss="modal"
aria-hidden="true" onclick="close1()">×</button>
<h4 class="modal-title">新建群组</h4>
<div class="modal-body">
<div class="form-group">
<label class="col-lg-2 col-md-2  control-label ">群组名</label>
```

```
<div class="col-md-8 controls ">
<input type="text"  id="name"  class="form-control" ></div>
<div class="clearfix"></div></div></div>
<div class="modal-footer">
<button type="button" class="btn btn-link" data-dismiss="modal"
                               onclick="close1()">取消</button>
<button  type="submit"class="btn  btn-primary"  onclick="savegroup(name)">保存
</button></div></div></div></div></div>
```

② 为“新建群组”按钮绑定单击事件“addgroup()”，显示对话框。

```
function addgroup() {
$("#groupform").modal("show");
}
```

③ 实现新建群组功能。

```
function savegroup() {
 var name = document.getElementById("name").value;
 if (name == null || name == ""){
  alert("请输入群组名");
  return false;
 }
if (name !== null && name != "") {
    var groupname = name;
    var data = {"name" : groupname};
    $.ajax({
      url : "addGroup.action",
      type : "post",
     dataType : 'json',
     data : data,
     success : function(s) {
      if (s.success) {
        location.reload();
        alert("创建成功");
        } else {
        location.reload();
        alert("创建成功");
        };
        }
     });
    }
}
```

### 5. 功能验证测试

将项目部署到 Tomcat 中并发布，项目运行成功后，登录进入，单击菜单栏的“群分享”按钮，获得群分享列表页，运行效果如图 8-12 所示。

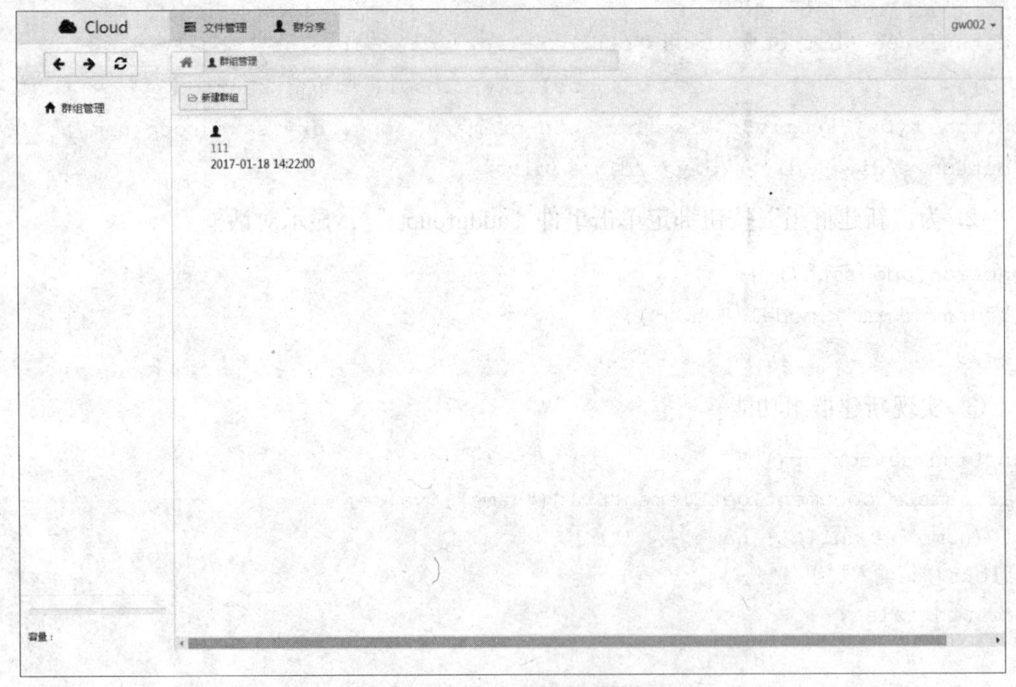


图 8-12 群分享功能运行效果图

本功能的测试场景见表 8-5。

表 8-5 群分享功能测试场景

| 编号 | 测试场景 | 输入参数 | 预期结果 |
|---|---|---|---|
| 1 | 单击新建群组的按钮 | 群组名 | 新建一个群组 |

## 任务 8.5 开发回收站功能

单击文件列表中的复选框，选择要删除的文件，单击“删除”按钮可以将文件删除至回收站中，在回收站中可以彻底删除文件。

### 8.5.1 相关知识

本功能实现的技术原理如下。

（1）视图层：显示回收站界面，如图 8-13 所示。

（2）控制层：接收视图层的消息，向服务层发送文件删除参数，并将服务层返回的结果发送到视图层。

（3）服务层：调用 Swift 接口删除文件，并将处理结果返回给控制层。

本功能的具体实现流程如图 8-14 所示。

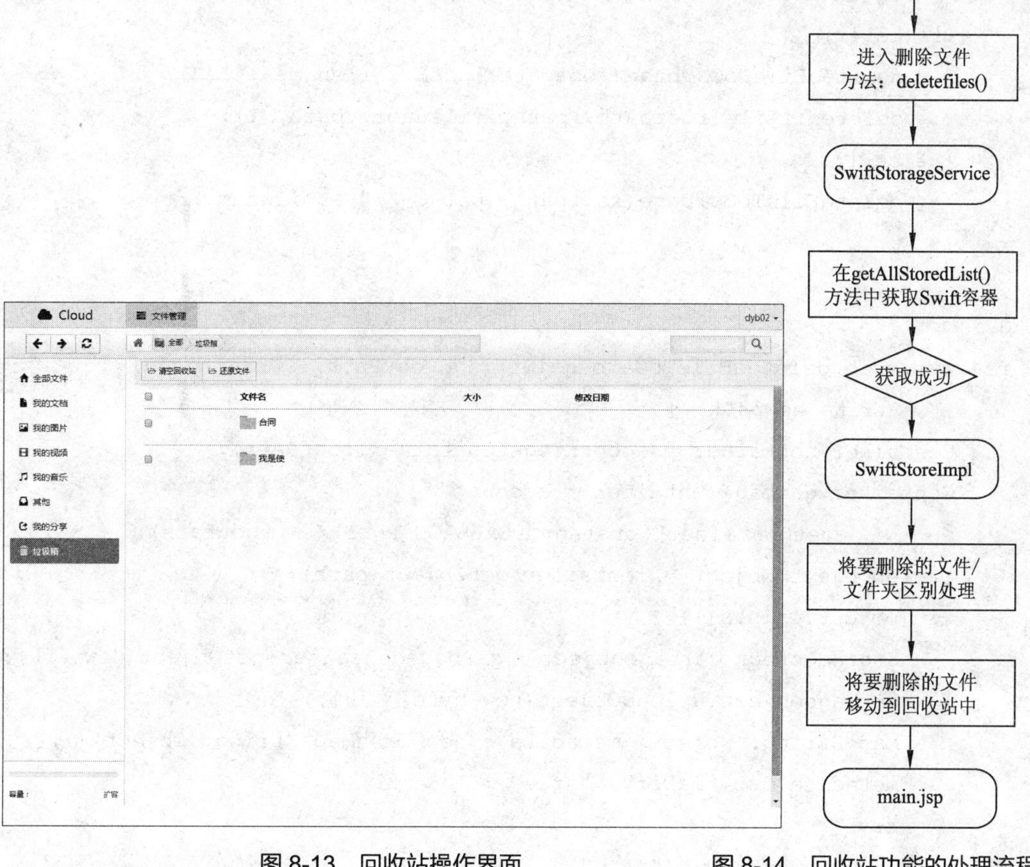


图 8-13　回收站操作界面　　　　图 8-14　回收站功能的处理流程图

## 8.5.2　实现步骤

### 1. 项目导入

在 Eclipse 中导入工程“project85”。

### 2. 实现服务层

（1）在“src/com/xiandian/cloud/storage/sh/SwiftStoreImpl.java”中增加如下代码：

```
/**
  * 描述：删除文件到垃圾箱
  *
  * @param rootpath：根路径
  * @param path：文件路径
  * @param filename：文件名称
  * @param isDir：是否是目录 true 是目录 false 不是目录
  */
  public void deletefiles(String rootpath, String path, String filename,
```

```
        boolean isDir) {
    // logger(rootpath+"执行 deletefiles 方法，删除文件到垃圾箱");
    if (isDir) {
        deletedFileToGarbage(rootpath, path, filename + "/");
        deletedirfile(rootpath, path, filename, path, 1);
    } else {
        deleteFileToGarbage(rootpath, path, "/" + filename);
    }

}
private void deletedFileToGarbage(String rootPath, String path,
        String newpath) {
    Container container = account.getContainer(rootPath);
    Container garbagecontainer = account
            .getContainer(Constants.GARBAGE_PREFIX + rootPath);
    StoredObject object = container.getObject(path);
    if (object.exists()) {
        StoredObject garbageobject = garbagecontainer.getObject(newpath);
        garbageobject.uploadObject(new byte[] {});
        Map<String, Object> metadata = new HashMap<String, Object>();
        metadata.put("isDir", "true");
        try {
            metadata.put("path",
                    URLEncoder.encode(object.getName(), "utf-8"));
        } catch (UnsupportedEncodingException e) {
            e.printStackTrace();
        }
        garbageobject.setMetadata(metadata);
        object.delete();
    }
}
/**
   * 删除文件夹到垃圾箱
   *
   * @param rootPath：根路径
   * @param path：文件路径
   * @param name：文件名称
   * @param epath：文件路径
   * @param type：值为1
```

```
     */
    private void deletedirfile(String rootPath, String path, String name,
            String epath, int type) {
        Container container = account.getContainer(rootPath);
        Container garbagecontainer = account
                .getContainer(Constants.GARBAGE_PREFIX + rootPath);
        Directory directory = new Directory(epath, '/');
        Collection<DirectoryOrObject> objects = container
                .listDirectory(directory);
        for (DirectoryOrObject c : objects) {
            if (c.isDirectory()) {
                String cpath = c.getAsDirectory().getName();
                String tpath = cpath.substring(path.length());
                String tnewpath = name + "/" + tpath;// strs[1];
                deletedFileToGarbage(rootPath, cpath, tnewpath);
                deletedirfile(rootPath, path, name, cpath, 2);
            } else {
                String temppath = c.getName();
                if (temppath.endsWith("/")) {
                    continue;
                }
                String tpath = temppath.substring(path.length());
                String tnewpath = name + "/" + tpath;// strs[1];
                deleteFileToGarbage(rootPath, temppath, tnewpath);
            }
        }
    }
    /**
     * 描述：删除文件到垃圾箱
     * @param rootPath 容器名称
     * @param path 删除文件的路径
     * @param newpath 文件名称当前路径
     * @return void
     */
    private void deleteFileToGarbage(String rootPath, String path,
            String newpath) {
    //  account = UtilTools.getAccount();
        Container container = account.getContainer(rootPath);
        StoredObject object = getObject(container, path);
```

```
        Container garbagecontainer = account
                .getContainer(Constants.GARBAGE_PREFIX + rootPath);
        StoredObject garbageobject = getObject(garbagecontainer, newpath);
        object.copyObject(garbagecontainer, garbageobject);
        Map<String, Object> metadata = new HashMap<String, Object>();
        metadata.put("isDir", "false");
        try {
            metadata.put("path", URLEncoder.encode(object.getName(), "utf-8"));
        } catch (UnsupportedEncodingException e) {
            e.printStackTrace();
        }
        garbageobject.setMetadata(metadata);
        delete(object);
    }
```

（2）在“src/com/xiandian/cloud/storage/service/SwiftStorageService.java”中新增如下代码：

```
/**
 * 取得垃圾箱路径下的所有文件
 */
public List getAllGarbageList(String username, String password) {
    Account account = SwiftUtilTools.getAccount(username, password);
    SwiftStoreImpl Swiftdfs = new SwiftStoreImpl(account);
    List<FileBean> list = Swiftdfs.getFile(Constants.GARBAGE_PREFIX
            + username);
    return list;
}
/**
 * 删除文件，移动到回收站，由于没有实现还原，所以简化实现删除
 *
 */
public void deleteFiles(String username, String password,
        List<Map<String, Object>> list) {
    Account account = SwiftUtilTools.getAccount(username, password);
    SwiftStoreImpl Swiftdfs = new SwiftStoreImpl(account);
    for (Map<String, Object> map : list) {
        String name = (String) map.get("name");
        String path = (String) map.get("path");
        String tisDir = (String) map.get("isDir");
```

```
        boolean isDir = "true".equals(tisDir) ? true : false;
        Swiftdfs.deletefiles(username, path, name, isDir);
    }
}
```

3. 实现控制层

在"/src/com/xiandian/cloud/storage/Web/StorageController.java"中新增如下代码：

```
/**
     * 取得垃圾箱的文件
     *
     * @param request
     * @param response
     * @param path
     * @return
     */
    @RequestMapping("/garbage")
    public ModelAndView garbage(HttpServletRequest request,
            HttpServletResponse response) {
        ModelAndView view = new ModelAndView();
        User user = getSessionUser(request);
        List list = storageService.getAllGarbageList(user.getUsername(),
                user.getPassword());
        view.addObject("list", list);
        view.setViewName("/garbage");
        return view;
    }

    /**
     * 删除文件
     *
     * @param request
     * @param response
     * @param name
     */
    @RequestMapping("/deletefiles")
    @ResponseBody
    public Object deleteFiles(HttpServletRequest request,
            HttpServletResponse response,
            @RequestBody List<Map<String, Object>> list) {
```

```
        User user = getSessionUser(request);
        storageService.deleteFiles(user.getUsername(),    user.getPassword(),
list);
        return new MessageBean(true, Constants.SUCCESS_5);
    }
```

4．实现前后台数据交互

回收站功能分为两步实现，首先选择要删除的文件，将文件删除至回收站，然后编写回收站界面代码并显示所有删除的文件。

（1）删除文件至回收站。

① 用户选择某一文件时，触发单击事件“show()”，实现显示/隐藏删除按钮功能。

```
function show() {
 var objTable = document.getElementById("tab");
 for (var i = 0; i < objTable.rows.length; i++) {
 var checkbox = objTable.rows[i].childNodes[1].childNodes[0].childNodes[5];
 if (checkbox.checked == true) {
 document.getElementById("rename").style.display = "block";
 document.getElementById("copy").style.display = "block";
 document.getElementById("download").style.display = "block";
 document.getElementById("share").style.display = "block";
 document.getElementById("delete").style.display = "block";
 break;
 }
 else {
 document.getElementById("rename").style.display = "none";
 document.getElementById("copy").style.display = "none";
 document.getElementById("download").style.display = "none";
 document.getElementById("share").style.display = "none";
 document.getElementById("delete").style.display = "none";
  }
}
```

② 给“删除”按钮绑定单击事件“deletefile()”，并实现删除功能。

```
//删除文件
function deletefile() {
if (confirm("你确定要删除所选文件吗？")) {
var objTable = document.getElementById("tab");
var data = '[';
for (var y = 0; y < objTable.rows.length; y++) {
```

```
var checkbox = objTable.rows[y].childNodes[1].childNodes[0].childNodes[5];
if (checkbox.checked == true) {
var path = checkbox.parentNode.parentNode.parentNode.childNodes[5].innerHTML;
path = decodeURIComponent(path);
var td3 = checkbox.parentNode.parentNode.parentNode.childNodes[3];
var imgpic = td3.getElementsByTagName("img");
if
(imgpic[0].src.substring(imgpic[0].src.indexOf("assets"))=="assets/images/Fo
lder.png") {
isDir = true;
var names = td3.childNodes[0].childNodes[1];
var name = names.innerText;
}
else {
isDir = false;
var names = td3.childNodes[0].childNodes[3];
var name = names.innerText;
}
data += '{"path":"' + path + '","name":"' + name+ '","isDir":"' + isDir + '"}';
if (y < (objTable.rows.length - 1)) {
 data += ',';
  }
 }
}
data += ']';
var data = eval('(' + data + ')');
$.ajax({
    url : "deletefiles.action",
    type : "post",
    contentType : "application/json; charset=utf-8",
    data : JSON.stringify(data),
    success : function(s) {
    if (s.success) {alert(s.msg);}
    location.reload();
    }
  });
 }
}
```

（2）实现回收站获取删除的文件。

```
<tbody id="tab">
 <c:forEach var="fb" items="${list}">
 <tr>
 <td style="top: -6px; position: relative; left: 13px;">
 <label class="checkbox table-checkboxposition" for="checkbox1">
 <span class="icons main-icons">
 <span class="first-icon fui-checkbox-unchecked"></span>
 <span class="second-icon fui-checkbox-checked"></span>
 </span>
 <input type="checkbox" name='check' class="main-tabinput" onclick="show()">
 </label>
 </td>
 <td><span style="display: block">
 <c:choose>
 <c:when test="${fb.isdirectory == true}">
 <img src="assets/images/Folder.png" class="objimg">
 <input name="objimg" type="text" style="display: none" value="${fb.name}">
 </c:when>
 <c:when test="${fn:contains(fb.name,'.txt') || fn:contains(fb.name,'.TXT')}">
 <img src="assets/images/Text.png" class="objimg">
 </c:when>
 <c:when test="${fn:contains(fb.name,'.docx')||fn:contains(fb.name,'.DOCX')}">
 <img src="assets/images/Word1.png" class="objimg">
 </c:when>
 <c:when test="${fn:contains(fb.name,'.doc')||fn:contains(fb.name,'.DOC')}">
 <img src="assets/images/Word1.png" class="objimg">
 </c:when>
 <c:when test="${fn:contains(fb.name,'.xls')||fn:contains(fb.name,'.xlsx')}">
 <img src="assets/images/Excel0.png" class="objimg">
 </c:when>
 <c:when test="${fn:contains(fb.name,'.XLSX')||fn:contains(fb.name,'.XLS')}">
 <img src="assets/images/Excel0.png" class="objimg">
 </c:when>
 <c:when test="${fn:contains(fb.name,'.PPT')||fn:contains(fb.name,'.PPTX')}">
 <img src="assets/images/PPT.png" class="objimg">
 </c:when>
```

```
<c:when test="${fn:contains(fb.name,'.pptx')||fn:contains(fb.name,'.ppt')}">
<img src="assets/images/PPT.png" class="objimg">
</c:when>
<c:when test="${fn:contains(fb.name,'.pdf') || fn:contains(fb.name,'.PDF')}">
<img src="assets/images/pdf.png" class="objimg">
</c:when>
<c:when test="${fn:contains(fb.name,'.mp3')||fn:contains(fb.name,'.MP3')}">
<img src="assets/images/Music.png" class="objimg">
</c:when>
<c:when test="${fn:contains(fb.name,'.cda')||fn:contains(fb.name,'.mid')}">
<img src="assets/images/Music.png" class="objimg">
</c:when>
<c:when test="${fn:contains(fb.name,'.wav')||fn:contains(fb.name,'.WAV')}">
<img src="assets/images/Music.png" class="objimg">
</c:when>
<c:when test="${fn:contains(fb.name,'.mp4')||fn:contains(fb.name,'.MP4')}">
<img src="assets/images/Video.png" class="objimg">
</c:when>
<c:when test="${fn:contains(fb.name,'.wmv')||fn:contains(fb.name,'.WMV')}">
<img src="assets/images/Video.png" class="objimg">
</c:when>
<c:when test="${fn:contains(fb.name,'.rmvb')||fn:contains(fb.name,'.RMVB')}">
<img src="assets/images/Video.png" class="objimg">
</c:when>
<c:when test="${fn:contains(fb.name,'.swf')||fn:contains(fb.name,'.SWF')}">
<img src="assets/images/Video.png" class="objimg">
</c:when>
<c:when test="${fn:contains(fb.name,'.flv')||fn:contains(fb.name,'.FLV')}">
<img src="assets/images/Video.png" class="objimg">
</c:when>
<c:when test="${fn:contains(fb.name,'.avi')||fn:contains(fb.name,'.AVI')}">
<img src="assets/images/Video.png" class="objimg">
</c:when>
<c:when test="${fn:contains(fb.name,'.rar')||fn:contains(fb.name,'.RAR')}">
<img src="assets/images/ZIP.png" class="objimg">
</c:when>
<c:when test="${fn:contains(fb.name,'.zip')||fn:contains(fb.name,'.ZIP')}">
```

```
<img src="assets/images/ZIP.png" class="objimg">
</c:when>
<c:when test="${fn:contains(fb.name,'.jpg') || fn:contains(fb.name,'.JPG')}">
<img src="assets/images/Picture.png" class="objimg">
</c:when>
<c:when test="${fn:contains(fb.name,'.png')||fn:contains(fb.name,'.PNG')}">
<img src="assets/images/Picture.png" class="objimg">
</c:when>
<c:when test="${fn:contains(fb.name,'.gif')||fn:contains(fb.name,'.GIF')}">
<img src="assets/images/Picture.png" class="objimg">
</c:when>
<c:when test="${fn:contains(fb.name,'.jpeg')||fn:contains(fb.name,'.JPEG')}">
<img src="assets/images/Picture.png" class="objimg">
</c:when>
<c:when test="${fn:contains(fb.name,'.ico')||fn:contains(fb.name,'.ICO')}">
<img src="assets/images/Picture.png" class="objimg">
</c:when>
<c:otherwise>
<img src="assets/images/other.png" class="objimg">
</c:otherwise>
</c:choose>
<span>${fb.name}</span>
</span>
<div class="edit-name" style="display: none;">
<input class="box" type="text" value="${fb.name}">
<a class="sure" href="javascript:void(0);"
    onclick="sure()"><span
    class="glyphicon glyphicon-ok"></span></a> <a
    class="cancel ml-10" onclick="cancel()"
    href="javascript:void(0);"><span
    class="glyphicon glyphicon-remove"></span></a>
</div></td>
<td class="table-path hide ta">${fb.path}</td>
<td>${fb.length}</td>
<td>${fb.lastmodified }</td>
</tr>
```

```
</c:forEach>
</tbody>
```

5. 功能验证测试

将项目部署到 Tomcat 中并发布，项目运行成功后，在文件列表中勾选需要删除的文件，单击菜单栏的“删除”按钮，删除的文件出现在回收站列表中，运行效果如图 8-15 所示。

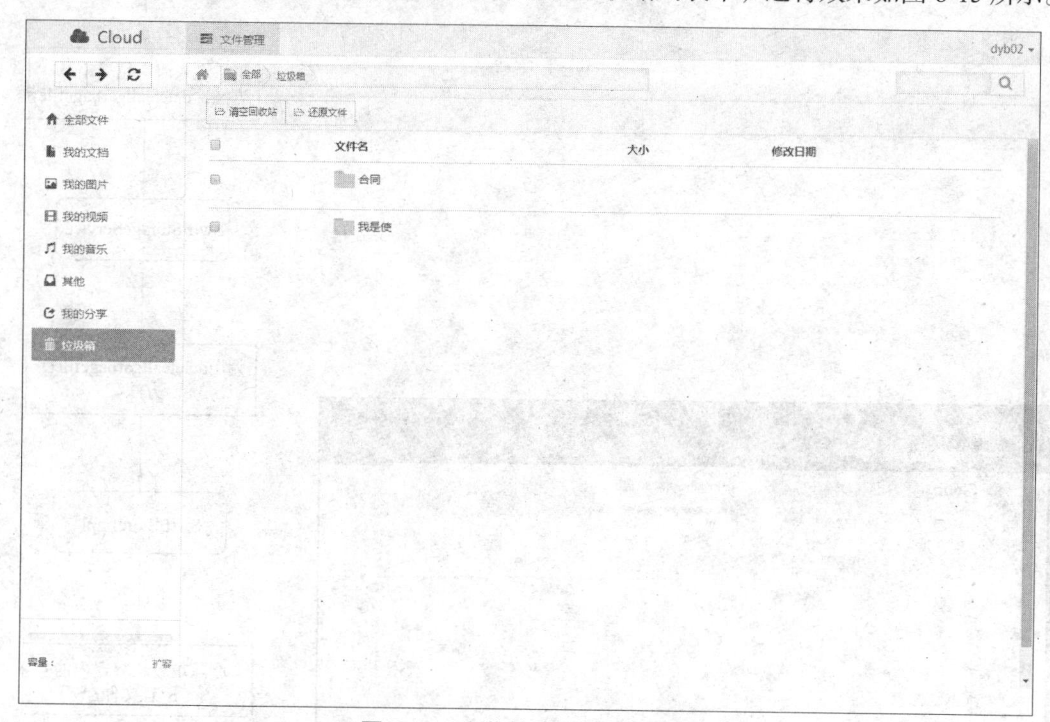


图 8-15 回收站功能运行效果图

本功能的测试场景见表 8-6。

表 8-6 回收站功能测试场景

| 编号 | 测试场景 | 输入参数 | 预期结果 |
|---|---|---|---|
| 1 | 登录成功后 | 删除文件 | 回收站显示被删除的文件 |
| 2 | 登录成功后 | 删除文件夹 | 回收站显示被删除的文件夹 |

## 任务 8.6 开发清空回收站功能

用户进入回收站页面，单击“清空回收站”按钮，程序将回收站内的所有文件都删除。

### 8.6.1 相关知识

本功能实现的技术原理如下。

（1）视图层：显示清空回收站的界面，如图 8-16 所示。

（2）控制层：接收视图层的消息，向服务层发送回收站删除参数，并将服务层返回的结果发送到视图层。

（3）服务层：调用 Swift 接口删除回收站中的文件，并将处理结果返回给控制层。

本功能的具体实现流程如图 8-17 所示。

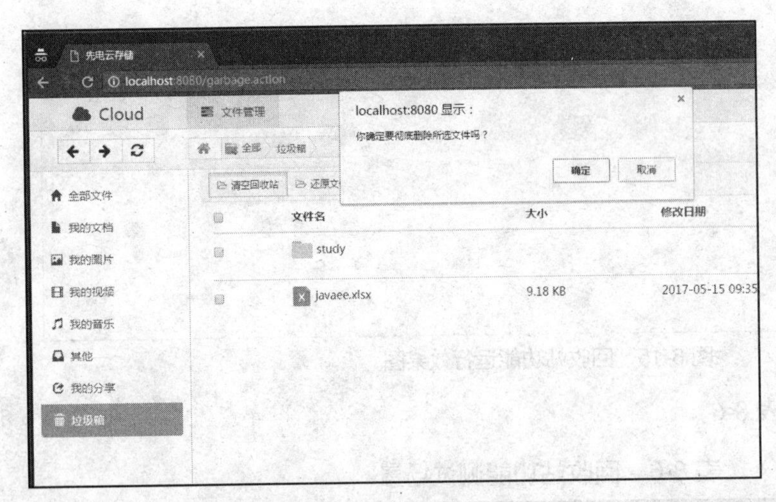


图 8-16　清空回收站操作界面

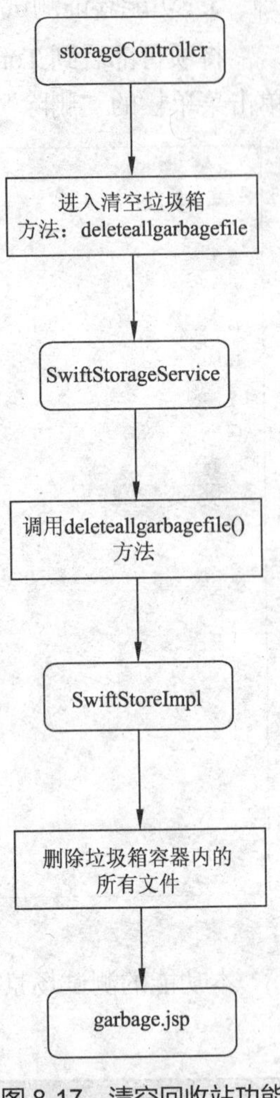


图 8-17　清空回收站功能的处理流程图

## 8.6.2　实现步骤

### 1. 项目导入

在 Eclipse 中导入工程“project86”。

### 2. 实现服务层

（1）在“src/com/xiandian/cloud/storage/service/SwiftStorageService.java”中增加如下方法。

```
/**
 * 清空回收站
```

```
 *
 * @param username
 * @param path
 */
public void deleteAllGarbageFile(String username, String password) {
    Account account = SwiftUtilTools.getAccount(username, password);
    SwiftStoreImpl Swiftdfs = new SwiftStoreImpl(account);
    Swiftdfs. deleteAllGarbageFile (username);
}
```

（2）在“src/com/xiandian/cloud/storage/sh/SwiftStoreImpl.java”中增加如下代码。

```
/**
 * 描述：删除回收站文件
 *
 * @param rootpath
 *            根路径
 */
public void deleteAllGarbageFile(String rootpath) {
Container container = account.getContainer(Constants.GARBAGE_PREFIX
            + rootpath);
    Collection<StoredObject> objects = list(container);
    for (StoredObject c : objects) {
        delete(c);
    }
}
```

3. 实现控制层

在“/src/com/xiandian/cloud/storage/Web/StorageController. java”中新增如下代码。

```
/**
 * 清空回收站的文件
 *
 * @param request
 * @param response
 * @param name
 */
@RequestMapping("/deleteallgarbagefile")
@ResponseBody
public Object deleteallGarbageFile(HttpServletRequest request,
        HttpServletResponse response) {
```

```
    User user = getSessionUser(request);
    storageService. deleteAllGarbageFile (user.getUsername(),user.getPassword());
    return new MessageBean(true, Constants.SUCCESS_5);
}
```

### 4. 实现前后台数据交互

为“清空回收站”按钮增加事件“deleteallgarbagefile()”，实现清空回收站功能。

```
//清空功能
function deleteallgarbagefile() {
        if (confirm("你确定要彻底删除所选文件吗？")) {
            $.ajax({
                url : "deleteallgarbagefile.action",
                type : "post",
                success : function(s) {
                    location.reload();
                }
            });
        }
    }
```

### 5. 功能验证测试

将项目部署到 Tomcat 中并发布，项目运行成功后，在文件列表中勾选需要删除的文件，单击菜单栏的“清空回收站”按钮，清空回收站内的所有文件，运行效果如图 8-18 所示。

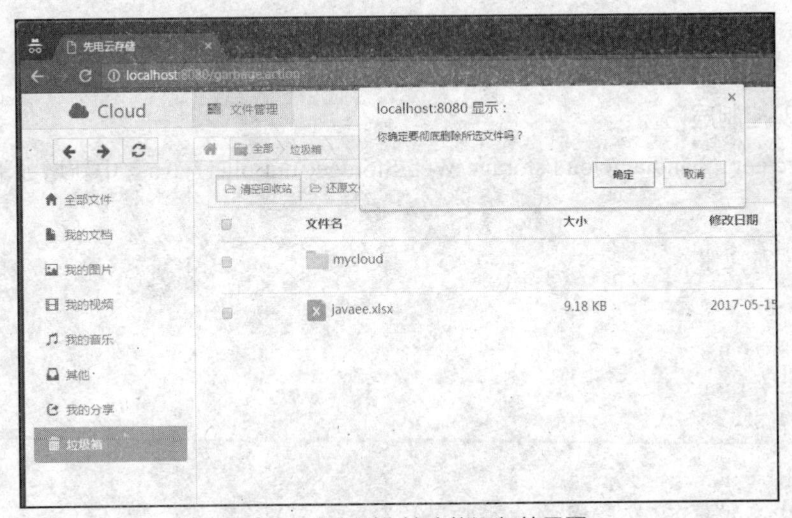


图 8-18　清空回收站功能运行效果图

本功能的测试场景见表 8-7。

表 8-7　清空回收站功能测试场景

| 编号 | 测试场景 | 输入参数 | 预期结果 |
| --- | --- | --- | --- |
| 1 | 进入回收站页面 | 选择清空回收站 | 回收站内所有文件删除 |

## 任务 8.7　开发还原文件功能

用户登录成功后，进入回收站页面，选择要还原的文件，单击“还原”按钮，将指定文件还原。

### 8.7.1　相关知识

本功能实现的技术原理如下。

（1）视图层：显示还原文件的界面，如图 8-19 所示。

（2）控制层：接收视图层的消息，向服务层发送文件还原参数，并将服务层返回的结果发送到视图层。

（3）服务层：调用 Swift 接口将指定文件还原到以前的位置，再将回收站内的文件删除，最后将处理结果返回给控制层。

本功能的具体实现流程如图 8-20 所示。

图 8-19　还原文件操作界面

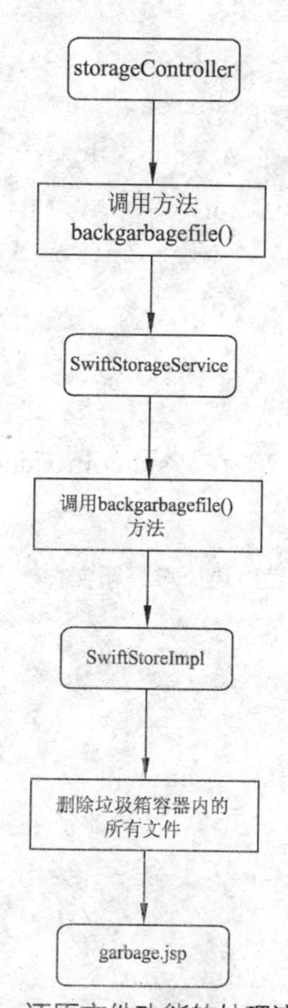


图 8-20　还原文件功能的处理流程图

### 8.7.2 实现步骤

#### 1. 项目导入

在 Eclipse 中导入工程“project87”。

#### 2. 实现服务层

（1）在“src/com/xiandian/cloud/storage/service/SwiftStorageService.java”中增加如下代码：

```
/**
 * 还原垃圾箱的文件
 *
 * @param email
 * @param path
 */
public void backGarbageFile(String username, String password,
        List<Map<String, Object>> list) {
    for (Map<String, Object> map : list) {
        String path = (String) map.get("path");
        String tisDir = (String) map.get("isDir");
        boolean isDir = "true".equals(tisDir) ? true : false;
        Account account = SwiftUtilTools.getAccount(username, password);
        SwiftStoreImpl Swiftdfs = new SwiftStoreImpl(account);
        Swiftdfs. backGarbageFile (username, path, isDir);


    }
}
```

（2）在“src/com/xiandian/cloud/storage/sh/SwiftStoreImpl.java”中增加如下代码：

```
/**
 * 描述：还原垃圾箱文件
 *
 * @param rootpath
 *            根路径
 * @param path
 *            文件路径
 * @param isDir
 *            true 目录 false 非目录
 * @return void
 */
public void backGarbageFile(String rootpath, String path, boolean isDir) {
```

```
    // logger(rootpath+"执行 backgarbagefile 方法，还原垃圾箱文件");
    if (isDir) {
        backdfile(rootpath, path);
        backdirfile(rootpath, path);
    } else {
        backfile(rootpath, path);
    }
}
private void backdfile(String rootpath, String path) {
    Container container = account.getContainer(rootpath);
    Container garbagecontainer = account
            .getContainer(Constants.GARBAGE_PREFIX + rootpath);
    StoredObject garbageobject = getObject(garbagecontainer, path);
    Map<String, Object> metadata = garbageobject.getMetadata();
    String oldpath = getOldPath(metadata);
    if (oldpath != null) {
        StoredObject object = getObject(container, oldpath);
        uploadObject(object);
        delete(garbageobject);
    }
}
private void backdirfile(String rootpath, String path) {
    Directory directory = new Directory(path, '/');
    Container garbagecontainer = account
            .getContainer(Constants.GARBAGE_PREFIX + rootpath);
    Collection<DirectoryOrObject> objects = garbagecontainer
            .listDirectory(directory);
    for (DirectoryOrObject c : objects) {
        if (c.isDirectory()) {
            String cpath = c.getAsDirectory().getName();
            backdfile(rootpath, cpath);
            backdirfile(rootpath, cpath);
        } else {
            String temppath = c.getName();
            if (temppath.endsWith("/")) {
                continue;
            }
            backfile(rootpath, temppath);
        }
```

```
    }
}
```

3. 实现控制层

在“/src/com/xiandian/cloud/storage/Web/StorageController. java”中增加如下代码：

```
/**
 * 还原文件
 *
 * @param request
 * @param response
 * @param name
 */
@RequestMapping("/backfiles")
@ResponseBody
public Object backfiles(HttpServletRequest request,
        HttpServletResponse response,
        @RequestBody List<Map<String, Object>> list) {
    User user = getSessionUser(request);
    storageService. backGarbageFile  (user.getUsername(),user.getPassword(),
list);
    return new MessageBean(true, Constants.SUCCESS_5);
}
```

4. 实现前后台数据交互

为“还原”按钮绑定单击事件“regarbagefile()”，实现还原功能。

```
//还原功能
function regarbagefile() {
if (confirm("你确定要还原所选文件吗？")) {
var objTable = document.getElementById("tab");
var data = '[';
for (var y = 0; y < objTable.rows.length; y++) {
var checkbox = objTable.rows[y].childNodes[1].childNodes[0].childNodes[5];
if (checkbox.checked == true) {
var checkselect = checkbox;
var filepaths = checkselect.parentNode.parentNode.parentNode.childNodes
var path = filepaths[7].innerHTML;
path = decodeURIComponent(path);
var td3 = filepaths[5];
var imgpic = td3.getElementsByTagName("img");
```

```
if (imgpic.length > 0) {
isDir = true;
var filename = td3.childNodes[0].childNodes[1].childNodes[0];
var name = $(filename).val();
} else {
isDir = false;
var filename = td3.childNodes[0].childNodes[0]
var name = $(filename).text();
}
data += '{"path":"' + path + '","name":"' + name+ '","isDir":"' + isDir + '"}';
if (y < (objTable.rows.length - 1)) {
data += ',';
 }
}
}
data += ']';
var data = eval('(' + data + ')');
$.ajax({
    url : "backfiles.action",
    type : "post",
    contentType : "application/json; charset=utf-8",
    data : JSON.stringify(data),
    success : function(s) {
    if (s.success) {
    alert("还原成功");
    } else {
    alert("还原失败");
    }
    location.reload();
    }
  });
 }
}
```

5. 功能验证测试

将项目部署到 Tomcat 中并发布，项目运行成功后，登录进入，在文件列表中勾选需要还原的文件，单击菜单栏的“还原文件”按钮，将选中的文件还原到原有的位置，如图 8-21 所示。

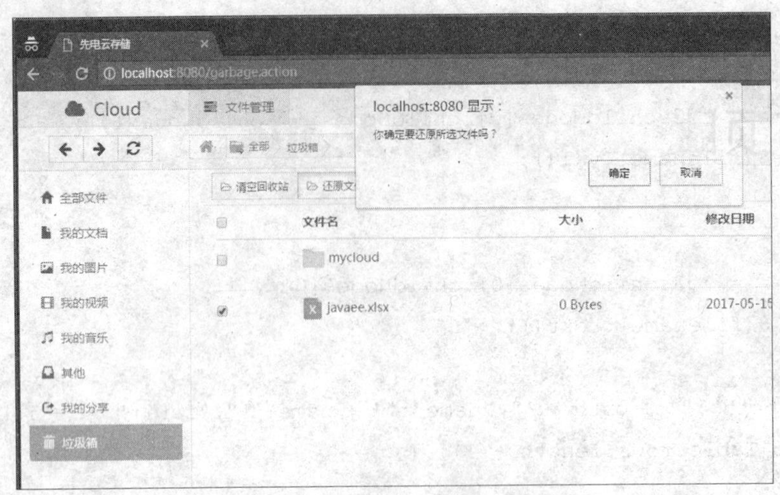


图 8-21　还原文件功能运行效果图

本功能的测试场景见表 8-8。

表 8-8　还原文件功能测试场景

| 编号 | 测试场景 | 输入参数 | 预期结果 |
|---|---|---|---|
| 1 | 进入回收站页面 | 还原文件 | 文件还原至原本位置 |
| 2 | 进入回收站页面 | 还原文件夹 | 文件夹还原至原本位置 |

# 项目 9 部署发布

## 单元介绍

本单元读者需掌握网盘的部署和发布流程。

## 学习任务

本单元主要完成以下学习目标：

- 掌握使用 Eclipse 生成项目 WAR 文件的操作步骤；
- 掌握在应用服务器中部署和发布项目的工作流程。

## 任务 9.1 软件部署

### 9.1.1 相关知识

本案例采用 Windows Server 2008 部署，选择相应服务器，在服务器中进行环境的安装（JDK、Tomcat、MySQL），并配置相应的环境变量。安装及配置方法可参照项目 2。

### 9.1.2 实现步骤

#### 1. 导入数据库

导入用户表数据库文件“user.sql”。

（1）选择“文件→创建连接”菜单命令，创建数据库连接，如图 9-1 所示。

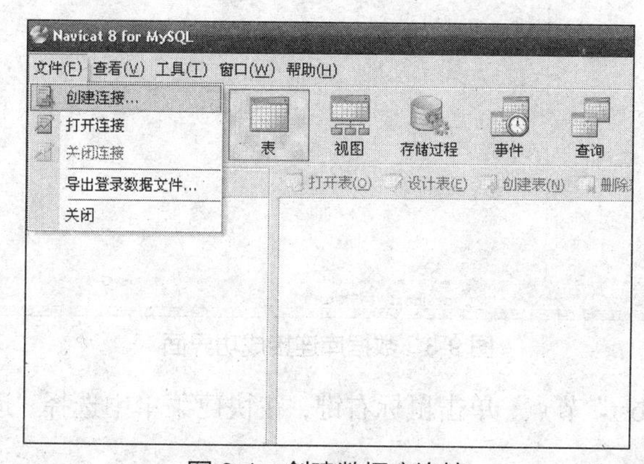


图 9-1 创建数据库连接

（2）输入数据库连接的相应信息，如图 9-2 所示。

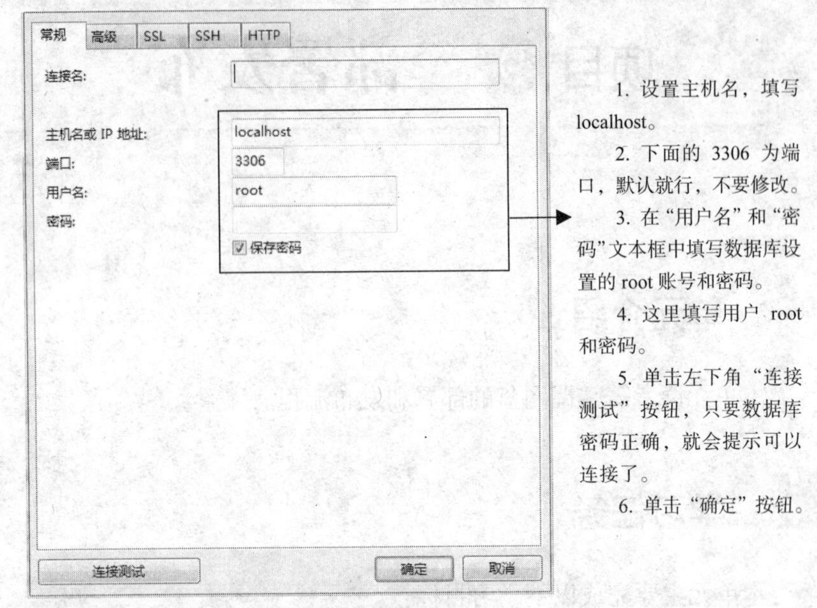


1. 设置主机名，填写 localhost。
2. 下面的 3306 为端口，默认就行，不要修改。
3. 在“用户名”和“密码”文本框中填写数据库设置的 root 账号和密码。
4. 这里填写用户 root 和密码。
5. 单击左下角“连接测试”按钮，只要数据库密码正确，就会提示可以连接了。
6. 单击“确定”按钮。

图 9-2　填写数据库连接参数

（3）数据库连接成功后，显示主界面，如图 9-3 所示。

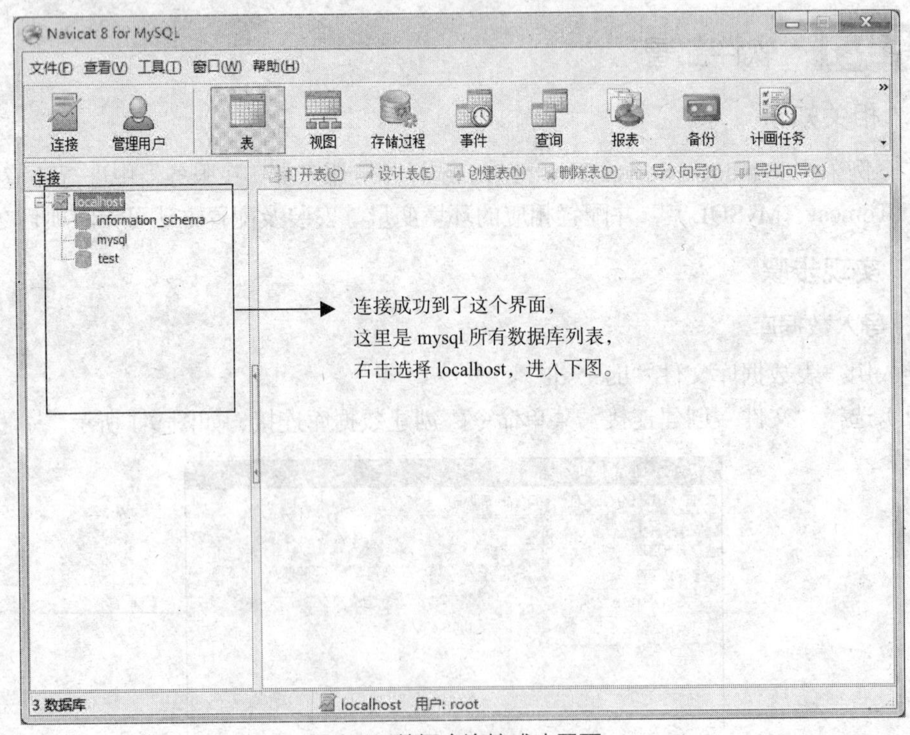


图 9-3　数据库连接成功界面

（4）选中“localhost”节点，单击鼠标右键，在快捷菜单中选择“运行批次任务文件”，如图 9-4 所示。

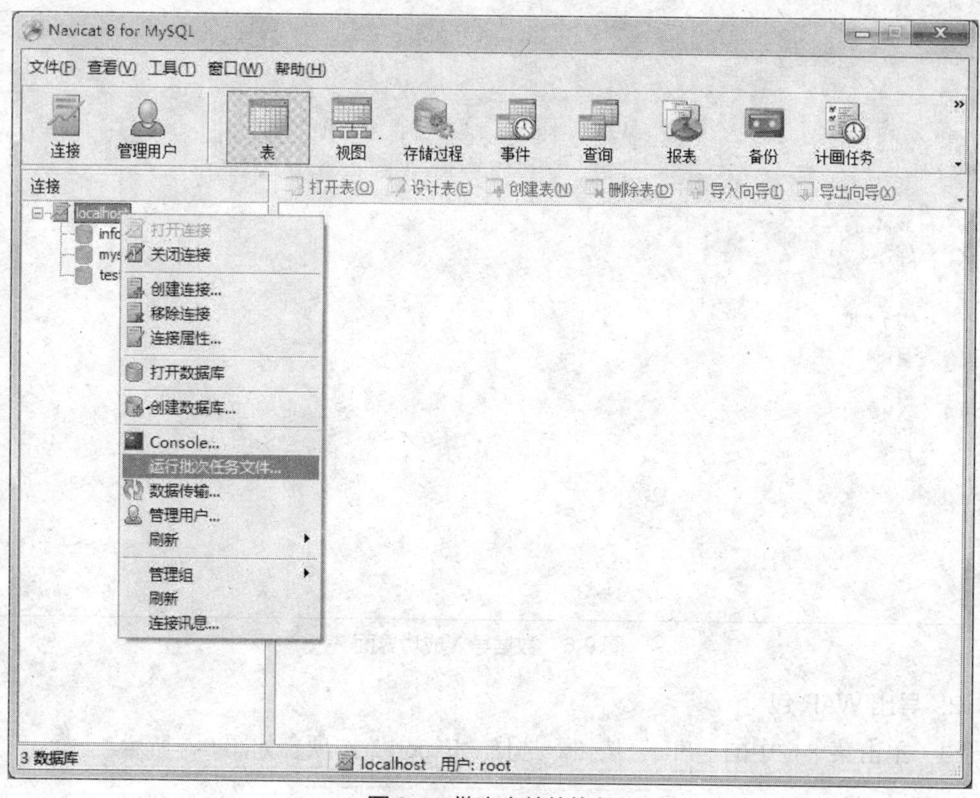


图 9-4　批次文件的执行

（5）解压“cloudstorage_web.zip”文件，在“cloudstorage_web\src\sql”目录下选择“user.sql”文件，单击“开始”按钮进行数据导入，如图 9-5 所示。

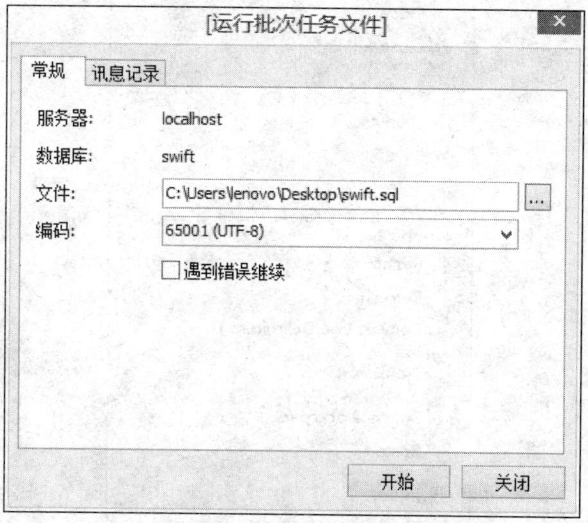

[运行批次任务文件]
常规　讯息记录
服务器:　localhost
数据库:　swift
文件:　C:\Users\lenovo\Desktop\swift.sql
编码:　65001 (UTF-8)
遇到错误继续
开始　关闭

图 9-5　导入数据到数据库

（6）成功后显示如下界面，如图 9-6 所示。

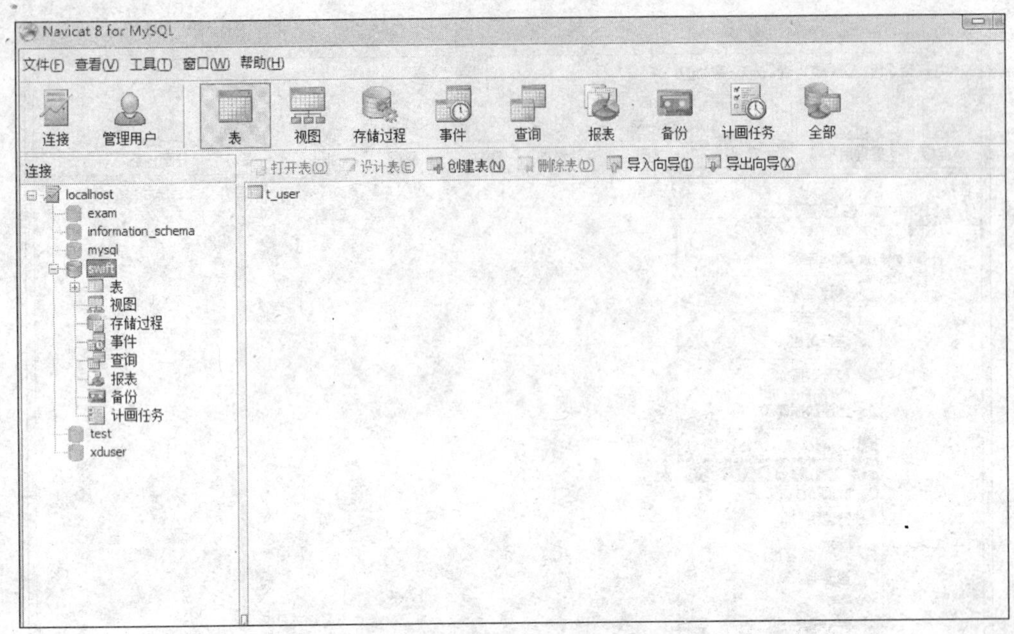


图 9-6 数据导入成功界面

## 2. 导出 WAR 包

（1）单击菜单栏中的“File”按钮，选择“Export”选项，如图 9-7 所示。

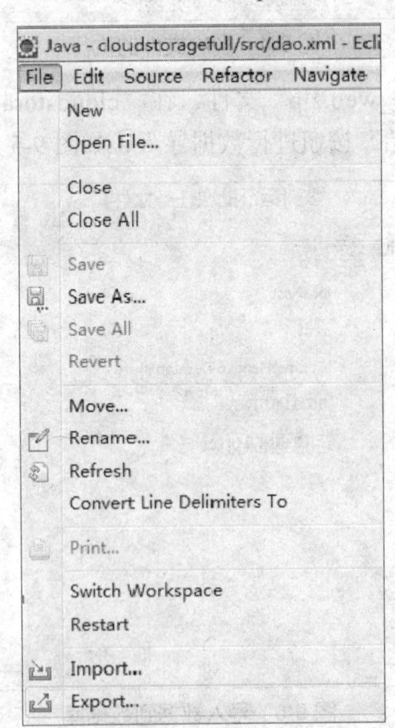


图 9-7 选择导出

（2）在弹出的窗口中选择“WAR file”，创建导出 WAR 类型，选定后单击“Next”按钮，如图 9-8 所示。

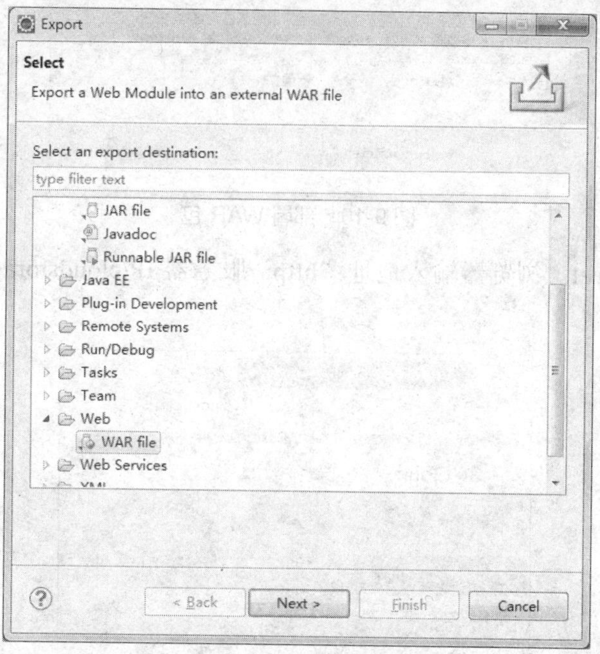


图 9-8　选择 WAR file

（3）在弹出的窗口中单击“Browse...”按钮确定导出路径，如图 9-9 所示。

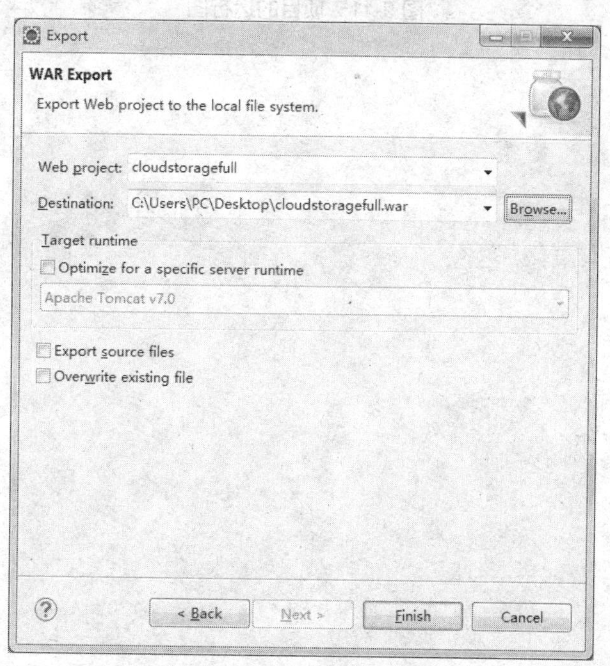


图 9-9　设置 WAR 包路径

（4）选择完成后单击“Finish”按钮，导出 WAR 包。

3．部署 WAR 包

（1）将 WAR 文件复制到 Tomcat 目录“\webapps\”下，如图 9-10 所示。

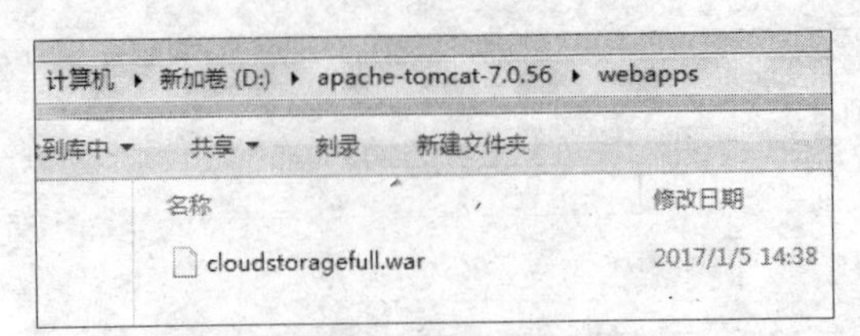


图 9-10 部署 WAR 包

（2）启动 Tomcat。浏览器输入地址“http://服务器 IP/cloudstoragefull”，如图 9-11 所示。

图 9-11 项目的运行图